FEMINIST
INTERPRETATIONS
OF
JOHN RAWLS

RE-READING THE CANON

NANCY TUANA, GENERAL EDITOR

This series consists of edited collections of essays, some original and some previously published, offering feminist re-interpretations of the writings of major figures in the Western philosophical tradition. Devoted to the work of a single philosopher, each volume contains essays covering the full range of the philosopher's thought and representing the diversity of approaches now being used by feminist critics.

Already published:

Nancy Tuana, ed., *Feminist Interpretations of Plato* (1994)
Margaret Simons, ed., *Feminist Interpretations of Simone de Beauvoir* (1995)
Bonnie Honig, ed., *Feminist Interpretations of Hannah Arendt* (1995)
Patricia Jagentowicz Mills, ed., *Feminist Interpretations of G. W. F. Hegel* (1996)
Maria J. Falco, ed., *Feminist Interpretations of Mary Wollstonecraft* (1996)
Susan Hekman, ed., *Feminist Interpretations of Michel Foucault* (1996)
Nancy Holland, ed., *Feminist Interpretations of Jacques Derrida* (1997)
Robin May Schott, ed., *Feminist Interpretations of Immanuel Kant* (1997)
Céline Léon and Sylvia Walsh, eds., *Feminist Interpretations of Søren Kierkegaard* (1997)
Cynthia Freeland, ed., *Feminist Interpretations of Aristotle* (1998)
Kelly Oliver and Marilyn Pearsall, eds., *Feminist Interpretations of Friedrich Nietzsche* (1998)
Mimi Reisel Gladstein and Chris Matthew Sciabarra, eds., *Feminist Interpretations of Ayn Rand* (1999)
Susan Bordo, ed., *Feminist Interpretations of René Descartes* (1999)
Julien S. Murphy, ed., *Feminist Interpretations of Jean-Paul Sartre* (1999)
Anne Jaap Jacobson, ed., *Feminist Interpretations of David Hume* (2000)
Sarah Lucia Hoagland and Marilyn Frye, eds., *Feminist Interpretations of Mary Daly* (2000)
Tina Chanter, ed., *Feminist Interpretations of Emmanuel Levinas* (2001)
Nancy J. Holland and Patricia Huntington, eds., *Feminist Interpretations of Martin Heidegger* (2001)
Charlene Haddock Seigfried, ed., *Feminist Interpretations of John Dewey* (2001)
Naomi Scheman and Peg O'Connor, eds., *Feminist Interpretations of Ludwig Wittgenstein* (2002)
Lynda Lange, ed., *Feminist Interpretations of Jean-Jacques Rousseau* (2002)
Lorraine Code, ed., *Feminist Interpretations of Hans-Georg Gadamer* (2002)
Lynn Hankinson Nelson and Jack Nelson, eds., *Feminist Interpretations of W. V. Quine* (2003)
Maria J. Falco, ed., *Feminist Interpretations of Niccolò Machiavelli* (2004)
Renée J. Heberle, ed., *Feminist Interpretations of Theodor Adorno* (2006)
Dorothea Olkowski and Gail Weiss, eds., *Feminist Interpretations of Maurice Merleau-Ponty* (2006)
Nancy J. Hirschmann and Kirstie M. McClure, eds., *Feminist Interpretations of John Locke* (2007)
Penny A. Weiss and Loretta Kensinger, eds., *Feminist Interpretations of Emma Goldman* (2007)
Judith Chelius Stark, ed., *Feminist Interpretations of Augustine* (2007)
Jill Locke and Eileen Hunt Botting, eds., *Feminist Interpretations of Alexis de Tocqueville* (2008)
Moira Gatens, ed., *Feminist Interpretations of Benedict Spinoza* (2009)
Marianne Janack, ed., *Feminist Interpretations of Richard Rorty* (2010)
Maurice Hamington, ed., *Feminist Interpretations of Jane Addams* (2010)
Nancy J. Hirschmann and Joanne Wright, eds., *Feminist Interpretations of Thomas Hobbes* (2012)

FEMINIST INTERPRETATIONS OF JOHN RAWLS

EDITED BY
RUTH ABBEY

THE PENNSYLVANIA STATE UNIVERSITY PRESS
UNIVERSITY PARK, PENNSYLVANIA

An earlier version of chapter 1, "Radical Liberals, Reasonable Feminists: Reason, Power, and Objectivity in MacKinnon and Rawls" by Anthony Simon Laden, appeared in the *Journal of Political Philosophy* 11, no. 2 (2003): 133–52. © Blackwell Publishing, 2003. Reprinted by permission.

An earlier version of chapter 2 appeared in Lisa H. Schwartzman, *Challenging Liberalism: Feminism as Political Critique* (University Park: Pennsylvania State University Press, 2006). Reprinted by permission.

Library of Congress Cataloging-in-Publication Data

Feminist interpretations of John Rawls / edited by Ruth Abbey.
   p. cm — (Re-reading the canon)
Summary: "A collection of essays that explore the philosophy and political theory of John Rawls from a variety of feminist perspectives"—Provided by publisher.
Includes bibliographical references and index.
ISBN 978-0-271-06179-5 (cloth : alk. paper)
ISBN 978-0-271-06180-1 (pbk. : alk. paper)
1. Rawls, John, 1921–2002.
2. Feminist theory.
3. Justice.
4. Liberalism.
I. Abbey, Ruth, 1961– , editor of compilation.
II. Series: Re-reading the canon.

JC251.R32F46   2013
305.4201—dc23
2013018008

Copyright © 2013 The Pennsylvania State University

All rights reserved
Printed in the United States of America
Published by The Pennsylvania State University Press,
University Park, PA 16802-1003

The Pennsylvania State University Press is a member of the Association of American University Presses.

It is the policy of The Pennsylvania State University Press to use acid-free paper. Publications on uncoated stock satisfy the minimum requirements of American National Standard for Information Sciences—Permanence of Paper for Printed Library Material, ANSI Z39.48-1992.

Contents

Preface   vii
List of Abbreviations   xi

Introduction: Biography of a Bibliography: Three Decades of Feminist Response to Rawls   1
    Ruth Abbey

1 Radical Liberals, Reasonable Feminists: Reason, Power, and Objectivity in MacKinnon and Rawls   24
    Anthony Simon Laden

2 Feminism, Method, and Rawlsian Abstraction   40
    Lisa H. Schwartzman

3 Rereading Rawls on Self-Respect: Feminism, Family Law, and the Social Bases of Self-Respect   57
    Elizabeth Brake

4 "The Family as a Basic Institution": A Feminist Analysis of the Basic Structure as Subject   75
    Clare Chambers

5 Rawls, Freedom, and Disability: A Feminist Rereading   96
    Nancy J. Hirschmann

6 Rawls on International Justice   115
    Eileen Hunt Botting

7 Jean Hampton's Reworking of Rawls: Is "Feminist Contractarianism" Useful for Feminism?   133
    Janice Richardson

8 Liberal Feminism: Comprehensive and Political 150
   Amy R. Baehr

   References 167
   List of Contributors 177
   Index 179

Preface

Nancy Tuana

Take into your hands any history of philosophy text. You will find compiled therein the "classics" of modern philosophy. Since these texts are often designed for use in undergraduate classes, the editor is likely to offer an introduction in which the reader is informed that these selections represent the perennial questions of philosophy. The student is to assume that she or he is about to explore the timeless wisdom of the greatest minds of Western philosophy. No one calls attention to the fact that the philosophers are all men.

Though women are omitted from the canons of philosophy, these texts inscribe the nature of woman. Sometimes the philosopher speaks directly about woman, delineating her proper role, her abilities and inabilities, her desires. Other times the message is indirect—a passing remark hinting at women's emotionality, irrationality, unreliability.

This process of definition occurs in far more subtle ways when the central concepts of philosophy—reason and justice, those characteristics that are taken to define us as human—are associated with traits historically identified with masculinity. If the "man" of reason must learn to control or overcome traits identified as feminine—the body, the emotions, the passions—then the realm of rationality will be one reserved primarily for men,[1] with grudging entrance to those few women who are capable of transcending their femininity.

Feminist philosophers have begun to look critically at the canonized texts of philosophy and have concluded that the discourses of philosophy are not gender-neutral. Philosophical narratives do not offer a universal perspective but rather privilege some experiences and beliefs over others. These experiences and beliefs permeate all philosophical theories, whether they be aesthetic or epistemological, moral or metaphysical. Yet this fact has often been neglected by those studying the traditions of philosophy.

Given the history of canon formation in Western philosophy, the perspective most likely to be privileged is that of upper-class white males. Thus, to be fully aware of the impact of gender biases, it is imperative that we reread the canon with attention to the ways in which philosophers' assumptions concerning gender are embedded within their theories.

This series, Re-Reading the Canon, is designed to foster this process of reevaluation. Each volume offers feminist analyses of the theories of a selected philosopher. Since feminist philosophy is not monolithic in method or content, the essays are also selected to illustrate the variety of perspectives within feminist criticism and to highlight some of the controversies within feminist scholarship.

In this series, feminist lenses focus on the canonical texts of Western philosophy, both those authors who have been part of the traditional canon and those philosophers whose writings have more recently gained attention within the philosophical community. A glance at the list of volumes in the series reveals an immediate gender bias of the canon: Arendt, Aristotle, Beauvoir, Derrida, Descartes, Foucault, Hegel, Hume, Kant, Locke, Marx, Mill, Nietzsche, Plato, Rousseau, Wittgenstein, Wollstonecraft. All too few women are included, and those few who do appear have been added only recently. In creating this series, it is not my intention to rectify the gender bias of the current canon of philosophical thought. What is and is not included within the canon during a particular historical period is a result of many factors. Although no canonization of texts will include all philosophers, no canonization of texts that excludes all but a few women can offer an accurate representation of the history of the discipline, as women have been philosophers since the ancient period.[2]

I share with many feminist philosophers and other philosophers writing from the margins of philosophy the concern that the current canonization of philosophy be transformed. Although I do not accept the position that the current canon has been formed exclusively by power relations, I do believe that this canon represents only a selective history of the tradition. I share the view of Michael Bérubé that "canons are at once the location, the index, and the record of the struggle for cultural representation; like any other hegemonic formation, they must be continually reproduced anew and are continually contested" (Bérubé 1992, 4–5).

The process of canon transformation will require the recovery of "lost" texts and a careful examination of the reasons why such voices have been silenced. Along with the process of uncovering women's philosophical history, we must begin to analyze the impact of gender ideologies upon the process of canonization. This process of recovery and examination must occur in conjunction with careful attention to the concept of a canon of authorized texts. Are we to dispense with the notion of a tradition of excellence embodied in a canon of authorized texts? Or, rather than abandon the whole idea of a canon, do we instead encourage a reconstruction of a canon of those texts that inform a common culture?

This series is designed to contribute to this process of canon transformation by offering a rereading of the current philosophical canon. Such a rereading shifts our attention to the ways in which woman and the role of the feminine are constructed within the texts of philosophy. A question we must keep in front of us during this process of rereading is whether a philosopher's socially inherited prejudices concerning woman's nature and role are independent of her or his larger philosophical framework. In asking this question, attention must be paid to the ways in which the definitions of central philosophical concepts implicitly include or exclude gendered traits.

This type of reading strategy is not limited to the canon but can be applied to all texts. It is my desire that this series reveal the importance of this type of critical reading. Paying attention to the workings of gender within the texts of philosophy will make visible the complexities of the inscription of gender ideologies.

Notes

1. More properly, it is a realm reserved for a group of privileged males, since the texts also inscribe race and class biases that thereby omit certain males from participation.

2. Mary Ellen Waithe's multivolume work *A History of Women Philosophers* attests to this presence of women.

Abbreviations

IPRR	Rawls, "The Idea of Public Reason Revisited" (1999)
JGF	Okin, *Justice, Gender, and the Family* (1989)
JFR	Rawls, *Justice as Fairness: A Restatement* (2001)
LP	Rawls, *The Law of Peoples* (1999)
OP	original position
PL	Rawls, *Political Liberalism* (1993)
TJ	Rawls, *A Theory of Justice* (1971)

Introduction

Biography of a Bibliography: Three Decades of Feminist Response to Rawls

Ruth Abbey

There is a wide consensus that John Rawls is one of the major thinkers of the twentieth century in the Anglophone world.... Most of contemporary political philosophy has been nurtured by his seminal ideas and can be understood either as a follow-up or a criticism and reaction against them. Thus, no student or scholar of the discipline can ignore them. (Audard 2007, 1)

By providing a chronological overview of English-language feminist engagements with Rawls from A *Theory of Justice* (*TJ*) onward,[1] this "biography of a bibliography" displays the range of issues canvassed by feminist readers of Rawls as well as their wide disagreement about the value of his corpus for feminist purposes. As we shall see, feminist responses to Rawls's first articulation of his theory of justice were, for the most part, critical yet hopeful. This early debate coalesced around such themes as whether his principles of justice extend to family relations; what it requires for a family to be just; and what lessons about justice children receive within families. The consensus was that whereas Rawls failed to exploit the feminist potential of his theorizing, it was there for others to explicate. Feminist responses to Rawls's work after his turn to political liberalism are more numerous and more polarized, with one side arguing that this turn has largely stripped justice as fairness of its feminist potential, and the other finding political liberalism to be replete with resources for addressing feminist concerns. In this phase of the debate, Rawls's feminist interlocutors examine whether efforts to accommodate reasonable pluralism compromise and constrain the feminist potential of justice as fairness. In doing so, they ask what qualifies as a reasonable comprehensive doctrine and how Rawls configures the public-private demarcation. They consider what political liberalism's appeal to public reason offers feminists and ponder whether justice as fairness evinces any awareness of the ways in which girls and women can internalize views of themselves as subordinate to men, and whether Rawls provides any measures to combat such self-interpretations of lesser worth. This introduction's survey of feminist engagements with Rawls also sets the stage for introducing the eight chapters in the current volume, chapters that testify to the continuing ambivalence among feminist readers about the value of Rawls's work.

Responses to *TJ*

Feminist engagement with *TJ* began with Jane English's 1977 article "Justice Between Generations," which responds to Rawls's attempt to address the problem of intergenerational justice by making the parties in the original position (OP) heads of families rather than individuals. Encouraging them to consider the interests of the next generation gives these choosers a more than simply self-interested perspective focused exclusively on the present. However, in English's estimation, this solution to the problem of getting one generation to save for the benefit of the next creates other problems. It is, for instance, inconsistent with Rawls's wider theoretical approach, which strives to minimize motivational assumptions (English 1977, 92–93). Rawls also fails to indicate how strongly these choosers care about their offspring and so cannot address how much they are willing to sacrifice in the present for their children's benefit (94). English also fears that if thinking in the OP takes the family, rather than the individual, as its basic unit, Rawls could effectively endorse some utilitarian-style calculi about overall costs and benefits in a way that violates his insistence upon the importance of each individual. Yet such an insistence is a major rationale for his rejection of utilitarianism in the first place (94–96). Making parties in the OP heads of families rather than individuals also troubles Rawls's account of justice within the family (91) and could legitimate sexist principles in the OP (94). English likens Rawls to Hegel when it comes to the family, because he seems to depict the family as a sphere of love and affection rather than justice (95).

Deborah Kearns elaborates on this point, inferring that "Rawls assumes love and the family unit to be so natural that he excludes them from the scope of the principles of justice" (Kearns 1983, 36; cf. 38). Placing the family beyond justice claims makes the two principles of justice applicable to the public realm only (39). Rawls never defines what he means by the family, but he seems to assume that it is a natural grouping and that its heterosexual, nuclear form is its natural form (38). He seems, further, to accept that moral development differs according to gender (37), with male and female children growing up with different experiences and expectations based on gender (36, 40). For Kearns, a family lacking justice is not well equipped to inculcate a sense of justice in its children (36, 40, 41). She concludes that "Rawls's whole theory is thus flawed from its very inception. An unjust family structure cannot produce just citizens" (36).

Like English and Kearns, Karen Green fears that Rawls's depiction of those in the OP as heads of families betrays a traditional understanding of the public-private separation that could be inimical to feminism. He also neglects gender as a consideration that those in the OP might have as they imagine who they will be once the veil of ignorance is lifted (Green 1986, 28–29). Yet women who conceive, bear, and nurture children are necessarily less free than the men who father them (29–30). However, just as it restricts natural liberty, so raising children brings benefits such as the fulfillment that comes from an intimate relationship with another human and the ability to transmit one's culture and conception of the good to the next generation (30). Contractors in the OP would wish to take this asymmetry into account by restricting both men's liberty to remove themselves from the reproductive process immediately after the sex act and women's monopoly on the benefits of raising children. Evincing more hope than Kearns does about the feminist potential of Rawlsian thought, Green offers a reading of the first principle of justice that would not only accommodate this need "to choose principles of justice appropriate for regulating a shared responsibility for and enjoyment of members of the younger generation" (31) but would also lead to "a radical feminist restructuring of family relations" (32; cf. 35)

Susan Moller Okin brings Rawlsian thought to bear on the debate between the ethic of care and the ethic of justice, which had been ignited by Carol Gilligan's 1982 book *In a Different Voice* (Okin 1989b, 246–47). Rawls seems, prima facie, to belong to the "ethic of justice" camp, for his approach to moral psychology can easily be read as rationalist, individualist, and abstract (230). However, Okin discerns "a voice of responsibility, care and concern for others" (230; cf. 236–38, 245) at the center of his thinking about justice, insisting that "feelings such as empathy and benevolence are at the very foundation of his principles of justice" (238). The emotional intelligence needed in the OP includes a capacity for empathy, for imagination, for thinking about difference, a willingness to listen to other points of view, and an ability to consider others as one's moral equals as well as to show care and concern for them (246; cf. 244–45, 247–48).

With her 1989 book *Justice, Gender, and the Family* (JGF), Okin produced the most sustained feminist response to *TJ*. Offering a feminist appraisal of Rawls's work, chapter 5, "Justice as Fairness: For Whom?" (89–109), issues a mixed review. Okin's first complaint is that *TJ* gives no indication that "modern liberal society . . . is deeply and pervasively gender-structured" (89; cf. 91). Neglect of gender is also evident in Rawls's initial failure to stipulate that parties in the OP would be ignorant of their gender (91).[2] While Okin applauds Rawls for including the family in his theory of justice, he fails to develop this promising beginning into any reflection on what justice within the family requires. Whereas Kearns read Rawls to be removing the family from considerations of justice, Okin sees him as carelessly assuming the family to be just.[3] In doing so, he prematurely and illegitimately forecloses a crucial area of normative enquiry about what justice in the family looks like and requires (22, 94, 97, 108; cf. Okin 1989b, 237–38, 249). Okin echoes English's suspicion that a conventional view of the family has been smuggled into his theory, and reiterates her charge that the family is "opaque to claims of justice" (English 1977, 75, cited in *JGF*, 94; cf. 95; cf. Okin 1989b, 230–31, 235, 239).

On the plus side, however, Okin finds considerable feminist promise and tries to exploit this in ways that Rawls had not (*JGF*, 90). She maintains that "a consistent and wholehearted application of Rawls's liberal principles of justice can lead us to challenge fundamentally the gender system of our society" (89). In addition to the conception of justice as requiring both reason and emotion, discussed above, three encouraging aspects are (1) the location of the family in the basic structure, (2) the device of the OP, and (3) the depiction of the family as a major school of citizens' moral development. Rawls's explicit inclusion of the family in the basic structure suggests that the family should be examined and criticized in light of the demands of justice. Including the family in this structure also confounds the traditional liberal reading of the public-private distinction, for this theory of justice applies to public institutions and to the family, which traditionally has been deemed private (93, 96–97). Assuming the veil of ignorance should require those in the OP to consider what sort of family life is most compatible with a just society. If ignorant of their gender, and formulating principles of justice relevant to the family, they would craft a society that disadvantages no one on the basis of gender (101, 105). Finally, Rawls recognizes, officially at least, the central role the family plays in producing and reproducing a just society.

Agreeing with Okin, Linda McClain identifies the many ways in which Rawls thinks of individuals as connected to, and caring for and about, one another, such that "both the care and justice perspectives, and their norms of response and reciprocity" appear in his thinking (McClain 1991–92, 1204; cf. 1206–9, 1214, 1217–18). McClain does not find that Rawls operates with an implicitly masculine image of the atomistic individual who is disconnected from others, self-interested, and competitive, and who prioritizes rights over responsibilities. That charge is based on a misunderstanding of the OP, which is simply a thought experiment; it is not freighted with ontological commitments (1204–5). McClain further suggests that the dichotomies between rights and responsibilities, justice and care, and masculine and feminine experience that structure some of the feminist criticism of liberalism are not illuminating when interpreting Rawls (1215, 1217, 1263). She holds that Rawls's actual view of persons—that they are capable of assessing and revising their conceptions of the good—can be beneficial for women, allowing them to review "whether certain kinds of connection, at the familial or community level, are desirable and not oppressive" (1206).

Lara Trout finds the Rawlsian approach to justice wanting because it focuses on economic interests to the neglect of other types of interest that can be salient. Rawls does not acknowledge "that 'least advantaged' could entail more than one's economic starting point in society" (Trout 1994, 44). His conceptualization of the "relevant social position" therefore manifests "an insensitivity toward the proper representation of minorities and women" (39). Noneconomic interests "are grounded in a person's self-respect" (40), which makes Trout's charge that Rawls ignores them a powerful one, given that self-esteem is an important primary good for him (41). Trout's answer to her own question—"can justice as fairness accommodate diversity?"—is "yes it can, even though it currently does not." Like many of the first responses to *TJ*, Trout's critique is immanent: she criticizes Rawls for failing to give one of his own claims the theoretical attention it deserves.[4]

Responses to Political Liberalism

A major development in Rawls's thinking about justice in the wake of *TJ* was his distinction between comprehensive and political liberalisms and his insistence that justice as fairness was a purely political doctrine. Born of his greater attention to pluralism in a liberal society, this innovation manifested Rawls's ambition to make justice as fairness applicable and appealing to people holding a wide variety of worldviews and moral positions. The source of this pluralism in a liberal society is freedom: in a society that respects and promotes the basic civil and political freedoms, individuals inevitably generate different and ultimately conflicting worldviews and moral positions (*PL*, xviii). Any liberal approach to justice that fails to create maximum room for such pluralism is therefore deeply flawed. Thus, in a series of essays published after *TJ*, culminating in *PL* in 1993, Rawls carves out a form of liberalism that is an exclusively political doctrine. All justice as fairness needs is an overlapping consensus on the principles of right, leaving individuals free to live according to their own values in other areas, assuming that these positions are reasonable.

One of the first feminist responses to this shift came from John Exdell, who argued that these changes had major, and mostly adverse, effects on its feminist potential (Exdell 1994, 450, 461). A principal aim of political liberalism is to minimize controversy about the principles of justice, which makes Rawls's approach more conservative than would be acceptable to feminists who challenge many accepted beliefs about women's (and men's) roles and status. This conservatism is compounded by Rawls's ambition to ground the principles of political liberalism in the shared meanings of the public culture (459). The reach of political liberalism is limited, too, for it is formulated for the public-political realm, respecting pluralism by leaving citizens free to live according to their own beliefs and choices in the private sphere (442, 460). While this has the desired effect of making political liberalism attractive to religious fundamentalists, it offers no support to feminist attempts to make standards of justice applicable to the family (442) and to increase women's liberty, agency, autonomy, and equality. Exdell identifies a number of philosophical and policy questions where feminism and fundamentalism collide (445–47, 449) and suggests that political liberalism effectively, even if unwittingly, lends support to the latter rather than the former.

Exdell concedes that political liberalism might not be wholly bereft of feminist resources—Rawls requires that all members of the well-ordered society be educated about their rights and liberties, which includes the freedom to leave groups they find oppressive or uncongenial and to develop their own conception of the good. This creates some space for an education in girls' and women's equality and autonomy and could justify some feminist policies to promote women's independence (1994, 453–54). But at best this points to Rawls's ambivalence about how tightly the public can be separated from the private, and it is ultimately defeated by his conception of public reason, which relies upon values and beliefs widely shared by the citizenry (460).

Linda Hirshman answers the question "Is the original position inherently male-superior?" with a resounding yes. She expresses amazement at Rawls's ability to ignore feminist responses to his work in particular, and feminist scholarship about the social contract tradition more generally (Hirshman 1994, 1860–61). Hirshman is especially critical of his "amputation of the public sphere from the rest of life" (1861, 1865); his confinement of justice issues to the state (1862–63); and his "metaphysical assumption of a moral psychology of autonomous individualism" (1861, 1865, 1868). Although she does not cite Exdell's work, Hirshman also castigates *PL*'s attempt to accommodate religious doctrines, for she sees the revival of Christian fundamentalism in American society as driven by regressive views of race and gender (1864–65). Any approach to justice that focuses on appealing to such views has, in effect, to legitimate inegalitarian doctrines.

Conducting a Rawlsian thought experiment of her own, with some "tough-minded" Hobbesian realism thrown in (1877), Hirshman rethinks "the social-contract scenario from the distaff side" (1875). Implicitly endorsing Okin's suggestion that the OP offers a fruitful way of thinking about justice for women, Hirshman imagines how contractors behind the veil of ignorance would think, were it not tacitly accepted that they were, and would remain, men. Although she does not cite Green's work, she echoes its concern that, being physically smaller and weaker and rendered vulnerable by childbirth and nursing, females will be handicapped when contracts are imagined among equal and autonomous individuals. Hirshman identifies two major entailments of those in the OP imagining that they might be women once the veil was lifted. "First, the application of the principles of justice could not possibly be limited to the official coercive agencies of the state, but would penetrate to the most intimate human relationships. This is vital for feminist purposes because many of the threats to women's liberty, equality, and security, such as rape, sexual abuse, and sex discrimination in the work place, emanate from non-state contexts" (1875–76). Second, rational contractors who knew they might end up as women would not be willing to bargain as individuals but would insist upon collective action to compensate for their physical disadvantages (1868, 1873–74, 1877).

Rawls's shift to a doctrine of political liberalism provoked an ambivalent response from Okin (1994, 23). On the plus side is his express claim that his principles of justice cover questions of gender equality (PL, xxix). Yet this claim remains underdeveloped, and Okin wonders whether he is offering women merely formal or more substantive equality (1994, 25, 39–41). Moreover, several areas of concern outweigh this positive feature. The family appears to have been consigned to the private realm (26), and Okin wonders whether Rawls retains any concern with justice within it, noting the reduced attention accorded to the family's role in forging a just society (23, 26, 28, 33–34, 37).[5] Okin also fears that Rawls is willing to accommodate sexist doctrines under the rubric of reasonable comprehensive doctrines (29–31). Like Exdell, whose analysis she praises (23n1), Okin observes that Rawls's toleration of a wide range of comprehensive doctrines clashes with the promotion of gender equality (28). Tolerating comprehensive doctrines that advocate gender inequality is not a good training ground for citizens in a just society, for when it comes to the political realm, justice as fairness requires that women be treated as free and equal. The friction between a comprehensive view of men and women as unequal and a political doctrine premised on women's equality would be significant, with the message that children learn in the household conflicting directly with the stance they must endorse as citizens (29). Like Exdell, Okin finds hope in the fact that schools are required to provide civic education, exposing children to the idea of citizens as free and equal, for this could mitigate sexist lessons inculcated in the home (31–32). Overall, though, the feminist potential that Okin identified in TJ seems to have diminished radically in Rawls's turn to political liberalism (25).

Sharon Lloyd addresses some of Okin's criticisms of political liberalism. What makes a comprehensive doctrine reasonable is its unwillingness to use state power to impose its views on others, so a sexist or racist doctrine that did not seek to use the state in this way could qualify as reasonable (Lloyd 1994, 355–56; 1995, 1323).[6] Rawls also tolerates inegalitarian conceptions of marriage, so long as the marriage has been entered into, and can be left, voluntarily. Lloyd thus effectively concedes the validity of some of Okin's concerns about political liberalism. But when Okin suggests that Rawls is uninterested in the justice of the family, Lloyd doubts that she has correctly understood the relationship between the principles of justice and the basic structure. These principles are not designed to apply directly to each and every institution within that structure, but rather to the operations of the system as a whole, to the outcomes of the interactions among its constituent institutions. What matters is that the operations of the institutions of the basic structure, when taken together, comply with the principles of justice (1994, 358–59; 1995, 1327).

Lloyd contends that Rawls is, in the interests of pluralism, indifferent to the form the family takes: all that matters is that it effectively prepare children for membership in a just society (1994, 358–59; 1995, 1328). This does not leave him insouciant toward family dynamics, however, or toward all forms of injustice within the family (1994, 361; 1995, 1331). A family in which the rights of one member were violated by another would be unacceptable. This precludes one family member's selling another into slavery, depriving him or her of civil and political liberties, and assaulting and battering him or her (1994, 358; 1995, 1327). Children cannot be neglected or abused and must be educated in the rights and duties of citizenship (1994, 362; 1995, 1332). Additional requirements for the family to be just include the aforementioned freedom to marry; guarantees of equitable division of material assets and parental support upon divorce; publicly provided or subsidized child care; family leave and flexible working hours; comparable worth policies in the workplace; and affirmative action for women and their equal access to equally good jobs (1994, 361–62; 1995, 1331–32). Lloyd concludes that "all of these measures taken together still allow families to adopt unequal divisions of labor, and affirm sexist beliefs about natural hierarchy" (1994, 362; 1995, 1332). Being raised in such a family could damage girls' and women's sense of equality and their capacity for justice, and this should pose a problem for Rawls (1994, 363–66; 1995, 1333–38). Lloyd suggests, however, that people can acquire a sense of gender justice despite being raised in households that do not practice or promote gender equality (1994, 364–65, 369; 1995, 1336, 1341–42).[7]

Kimberly Yuracko would agree with Lloyd's interpretation of Rawls's tolerance of inegalitarian gender doctrines in the family. But as the second half of her article's title implies, she is very critical of this. Yuracko's "Towards Feminist Perfectionism: A Radical Critique of Rawlsian Liberalism" (1995) objects that some of the comprehensive doctrines that Rawls would accept as reasonable encourage girls and women to think of themselves as subordinate. Although the basic structure transcends the standard public-private divide by including the family, *PL* "only applies his principles [of justice] to the public sphere" (Yuracko 1995, 6). But this obscures what Yuracko calls the spillover effects from the public to the private realm. What happens to individuals in the domestic sphere shapes their opportunities in other areas. If some family members carry an undue burden of unpaid domestic work, this limits their ability to participate in politics or other activities (7–9; cf. 3). Spillover effects also occur in more subjective registers, such as women's preferences and self-esteem (3, 12). Yuracko effectively echoes Trout's claim that Rawls attends to socioeconomic disadvantage while ignoring other forms of disadvantage and exclusion (13). She evinces skepticism about the sufficiency of his insistence that all children receive a civic education in their rights and duties as citizens, fearing that this is unlikely to go far enough in combating the effects of socialization in gender inequality (10–11, 13, 17–18, 32–35). Rawlsian liberalism is incapable of addressing the serious barrier to gender equality created by "internalized gender-based conceptions of the self" (2). Rather than accommodate the maximum number of conflicting comprehensive doctrines, a feminist approach to justice requires "an affirmative endorsement of the values and behaviors that are necessary for a normative vision of gender equality" (3; cf. 4, 31, 41–42, 48).

Amy Baehr (1996, 49–53) echoes Lloyd's observation that Okin has misunderstood the role that Rawls assigns to the two principles of justice. Baehr proposes that it would be possible to make the family a more just institution without requiring the direct application of the two principles, and she points out that whatever Okin's professed belief on this matter, she does not consistently apply the two principles to family life (61). Baehr, like Yuracko, would agree with Lloyd's interpretation of Rawls's tolerance of inegalitarian gender doctrines in the family under the mantle of the reasonable (57–58). This is one of the reasons why Baehr also agrees with Exdell and Okin about the diminution in feminist potential with Rawls's shift to political liberalism: he "dulls the critical edge of liberalism by capitulating too much to those holding sexist doctrines" (49; cf. 50, 56–57, 61). Baehr rehearses Yuracko's point about spillover effects (which she calls seepage) between private and public spheres making any strong separation between them untenable (51, 54, 58). She concludes by tendering a Habermasian approach to the public and the private as more fruitful for feminist purposes, for Habermas deems personal and political autonomy to be "co-original," positing that rights in the public sphere enjoy no normative priority over those in the private sphere (62).

Two angles from which the question of family justice can be considered within Rawls's work are discerned by Véronique Munoz-Dardé. The first is justice within the family. This part of her discussion retreads much of the ground covered in English's early response to Rawls.[8] Munoz-Dardé engages Okin's views on the feminist potential of Rawls's work (1998, 345–47) but finds her interpretation of those in the OP practicing empathy, care, and concern for others to be problematic and even unnecessary: ignorance of their sex "is enough to obtain the result Okin desires, namely that gender equality be considered as an issue of justice" (346). The second angle scrutinizes the justice of the family as an institution. Rawls poses the radical question of whether any family form is compatible with justice, for the family compromises equal opportunity. Because belonging to one family rather than another affects children's life prospects in myriad ways, "*as long as the family exists* individuals will have unequal life chances" (339, emphasis in original; cf. 335, 338). But Rawls argues against the family's abolition on the grounds that the family is the most effective way of ensuring children's moral development (339, 350–51). Unlike previous commentators, Munoz-Dardé treats *TJ* and *PL* as a single theory (335n1). She admits, however, that there have been changes (337), such as Rawls's seeming to adopt a more expansive conception of the family (338) and adding gender to the slate of informational restrictions in the OP (340). Munoz-Dardé cannot see how Okin's ideal of a liberalized family is compatible with political liberalism because it seems to rest upon a wider comprehensive doctrine, but here she fails to acknowledge that Okin's ideal was connected with *TJ* and that Okin herself expressed grave doubts about whether political liberalism could support it (347).

The justice of the family is also taken up by Ron Mallon, who agrees that the family is at odds with Rawls's commitment in the second principle of justice to equality of opportunity (Mallon 1999, 273). Arguing that the family cannot be successfully defended in terms of its unique contribution to children's moral development (275–81), Mallon advances an alternative defense based on the good of cultural membership (271). Although Rawls does not make cultural membership a primary good, he should, according to Mallon (272, 275, 281). Mallon sets out to show that (1) the family is a necessary and not merely a sufficient condition for the transmission of cultural membership, and (2) the primary good of cultural membership is of sufficient weight to override the goods ensured by the principle of fair equality of opportunity (281, 284–85). Given *PL*'s concern with a plurality of conceptions of the good, culture has to assume a significant role, for cultures host such conceptions as well as providing normative and symbolic resources for their revision (288). Cultural membership is therefore vital to the power of personhood that Rawls associates with a capacity for a conception of the good. Rawls's restrictions on the state's promoting any particular conception mean that the state could not carry out the task of transmitting cultural membership to children (289). Consistent with Rawls's emphasis on pluralism in *PL*, Mallon permits a wide variety of family forms to carry out this crucial task (271, 290–91).

An influential critique of Rawlsian liberalism inspired by the ethic-of-care debate was mounted by Eva Feder Kittay, who charges that its image of the liberal subject as an independent individual standing in a condition of equality relative to similar others blinds Rawls's political theory to the pervasive fact of human dependency. As children, we all experience dependency, but it recurs for shorter or more protracted periods through illness, mental or physical disability, financial need, and old age. This fact of dependency entails the work of caring for those who are dependent. Yet this social reality is occluded in justice as fairness, rendering Rawls unable to address the fact that most of the caring work in Western societies—care in the home for dependent family members, be they children, older people, or the disabled, and caring work in hospitals and other health institutions, child-care centers, and so on—is done by women. It is mostly unpaid or underpaid and always underrecognized (Kittay 1999, 78, 110–11). Any theory of justice unable to acknowledge, let alone accommodate, the fact of dependency and its implications for those who need, as well as those who provide, care, is seriously flawed. Kittay insists that dependency "must be faced from the beginning of any project in egalitarian theory that hopes to include *all* persons within its scope" (77, emphasis in original). Among the options for making justice as fairness more attentive to the fact of dependency are including the capacity to care as a moral power; considering dependents and their caregivers among the least well off; and adding a third principle of justice: *"To each according to his or her need for care, from each according to his or her capacity for care, and such support from social institutions as to make available resources and opportunities to those providing care"* (113, emphasis in original).

Just as Okin complained that major contemporary theorists of justice were largely oblivious to gender issues (*JGF*, vii, 14, 134), and as Kittay accentuates dependency relations, so Hilda Bojer observes that most contemporary theories of distributive justice ignore children (Bojer 2000, 23–24). She aims "to sketch a theoretical framework that makes it possible to take the just interests of children as children into account on an equal footing with those of adults" (26). Rawlsian theory is the only approach that can assist Bojer, but even then she must spell out its "radical and feminist" implications in a way that Rawls did not (30). If those behind the veil did not know whether they would be children or adults, the conditions of childhood would be taken very seriously, given how weak children are as a group. The important impact that childhood has on the adult that one becomes also urges that it be taken seriously by rational contractors (35). Although Bojer does not engage the debate about the family as an impediment to equal opportunity, one of her conclusions is compatible with that concern: "The right nutrition, preventive health care, education . . . must be ensured for every child in the just society, in my interpretation of Rawls, independently of their parents' income or merits" (37).

Samantha Brennan and Robert Noggle echo claims by English, Okin, and others regarding Rawls's failure to consider the justice of the family, and they emphasize how this neglects the interests of children (Brennan and Noggle 2000, 46). Rawls says little about children in *TJ* and even less in *PL*, they point out (47, 55, 61). When he does discuss children, it is in the context of moral development rather than their moral status in its own right (46). While this renders his theory of justice incomplete, Brennan and Noggle share Bojer's belief that Rawls's concern with the least advantaged members of society can be applied to children (46–47). Dropping the idea, in response to English's critique, that the parties in the OP are heads of families (*PL*, xlviiin20, 20n22, 274n12) enables the contractors to think of themselves as being children, once the veil of ignorance has lifted (Brennan and Noggle 2000, 52, 54), and therefore to construct principles of justice that protect children.

Stephen de Wijze reviews the feminist criticisms of political liberalism by Exdell and Okin and deems them unfounded. The first line of criticism, which Wijze calls "the incongruence argument," points to the mismatch between political liberalism's requirement that women be treated as free and equal citizens and its tolerance of sexist comprehensive doctrines in the household (Wijze 2000, 262–63). A second criticism, which Wijze dubs "the political virtues argument," asserts that a family in which injustice occurs is a poor training ground for citizens of a just society (263). Underpinning both lines of criticism is a concern with the way in which the liberal tradition has typically separated the public and private realms (263–65). Wijze contends that feminist critics misunderstand two key things: what political liberalism means by just family forms and how it draws the political-personal distinction (257). Following Lloyd and Baehr, he argues that Okin has misunderstood the relationship between the family and Rawlsian principles of justice, for the latter are not intended to govern the former directly (273–75, 278–79). Political liberalism operates with a robust but minimal conception of family justice, according to which a just family respects its members' civil rights and liberties and promotes their capacities to be or become cooperating members of society (280).

Even though the principles of justice do not apply directly to the family, they do constrain members' interactions with one another.

Wijze's second response is that feminist critics of political liberalism exaggerate the public-private separation. Political liberalism does not depict the family "as an inviolate inner sanctum immune from state interference" (272). Instead, as indicated above, civil rights and liberties are protected in the private sphere (274; cf. 275). In response to the political virtues argument, Wijze replies that if sexist family forms made it impossible for women "to develop the requisite political conception of the self" (277), they would not be tolerated by political liberalism. However, he charges that feminist claims about this are empirically dubious. Rather than provide evidence to counter these claims, he simply casts doubt on their plausibility. But even if sexist family forms prevent women from realizing the political conception of the self to the fullest possible degree, they will, in the interests of pluralism, continue to be acceptable (277).

Rawls's Later Writings

When considering the feminist implications of political liberalism, Lloyd made reference to an unpublished manuscript by Rawls from 1994 on women and the family (Lloyd 1994, 371n4). Some of the ideas in that manuscript made their way into section 5 of "The Idea of Public Reason Revisited," "On the Family as Part of the Basic Structure" (IPRR, 156–64). There Rawls mentions Lloyd along with McClain, Okin, and Martha Nussbaum as encouraging him "to think that a liberal account of equal justice for women is viable" (156–57n58). References to this later material begin to appear in feminist responses to Rawls in the twenty-first century.

In "Rawls and Feminism," Nussbaum conducts an overview of some of the main lines of feminist critique of justice as fairness, contending that Rawls can provide a powerful response to many (Nussbaum 2003, 488). Nussbaum devotes a lot of attention to feminist critiques of the rationalism and individualism of Rawlsian thought and endorses readings like those of Okin and McClain on the important role of emotions and connection with others in Rawlsian thinking (489–99). In Nussbaum's estimation, justice as fairness faces a greater challenge in incorporating women's equality into the family. On the one hand, Rawls's pluralist commitment to individuals' freedom to devise conceptions of the good implies maximum freedom for families. On the other, the family can be a site of "sex hierarchy, denial of equal opportunity, and also sex-based violence and humiliation"

(500). Rawls should acknowledge the liberal dilemma when it comes to the family, asking how to "balance adult freedom of association, and other important interests in pursuing one's own conception of the good, against the liberties and opportunities of children as future citizens" (506). Despite the more explicit and focused attention Rawls gave to women, the family, and justice in IPRR, Nussbaum identifies three remaining issues. First, if the family belongs to the basic structure, why does Rawls repeatedly liken it to voluntary associations (504)? Participation in a family is never voluntary for children, so freedoms of entry and exit hold little relevance for them. Second, Rawls gives insufficient weight to the fact that the family is a product of the law. The law defines which combinations of individuals can be regarded as families for the purposes of public policy, and shapes the nature of many of their relationships to one another and to other institutions. Because the state has the power to dictate who and what counts as a family, it is unhelpful to ponder the issue of state intervention in the family as if they were two discrete entities (504–6). Third, even though Rawls seems to have expanded his conception of the family since *TJ*, Nussbaum suspects that he endows the Western nuclear family with a quasi-natural status, ignoring forms of affiliation in other cultures that represent viable alternatives to the family (504).

Nussbaum also raises the issue, which goes back to Exdell and Okin, of political liberalism's tolerance of sexist comprehensive doctrines (2003, 509–11). She follows defenders like Lloyd and Wijze in pointing out that doctrines denying women's equality qua citizens would not be tolerated. She agrees that citizens who affirm women's subordination in the household will encounter friction between their views and political liberalism's insistence that women are equal citizens (510). Nussbaum also points out that feminists remain free to criticize such doctrines in civil society and personal relations (511). She concludes her review of feminist engagements with Rawls with Kittay's dependency critique (511–14), accepting that in order to respond persuasively, Rawls would have to rethink both his image of citizens in the well-ordered society as fully cooperating members over the course of a life and his account of primary goods. But, overall, Nussbaum believes that Rawls can productively absorb feminist criticisms by revising justice as fairness to become even more liberal (515).

A revised version of Anthony Laden's 2003 article "Radical Liberals, Reasonable Feminists: Reason, Power, and Objectivity in MacKinnon and Rawls" is published here as chapter 1. Laden questions the belief that political liberalism bodes poorly for feminists, showing instead that Rawlsian liberalism can engage the concerns of feminist critics of liberalism such as

Catharine MacKinnon. Laden distinguishes a theoretical approach that is blind to such concerns from one that is merely shortsighted. Liberalism would be blind to feminist concerns if there were nothing in its theoretical apparatus to respond to them. A theory that is merely myopic, by contrast, might overtly ignore such concerns, but it nonetheless contains resources for addressing them. Advancing a reading of Rawls's thought that emphasizes its deliberative dimensions, Laden sets out to show that political liberalism is merely shortsighted when it comes to gender.

Elizabeth Brake agrees with Laden that Rawlsian liberalism can rise to some of the challenges posed by MacKinnon's feminist critique of liberalism. Focusing on the issue of neutrality, Brake urges feminists to see it "as a moral ideal in the process of justification . . . that . . . will require substantive feminist reform" (Brake 2004, 295). The ideal of neutrality can be invoked to attack the sort of state-sponsored gender hierarchy of which MacKinnon is so critical (296). This initially seems counterintuitive, given that neutrality requires that the state refrain, as far as possible, from endorsing any conception of the good (296–98). But, as Brake points out, if current practices and institutions are nonneutral, the ideal of neutrality can leverage considerable change. Current arrangements regarding paid employment and unpaid caring work in the home are biased against women, so a neutral state would be justified in enacting many of the policy changes Okin outlines in *JGF*, in the name of removing or reducing this bias (307–9). Neutrality's appeal rests ultimately on the commitment to moral equality and equal respect that underlie it, for these values portend "a rejection of sexist and other oppressive doctrines" (309; cf. 301–2, 305)

In "Democratic Citizenship v. Patriarchy: A Feminist Perspective on Rawls," Marion Smiley reviews a number of feminist criticisms of Rawls, asking which, if any, prove fatal. The three most powerful are (1) that Rawls's methodology is abstract and individualist, which biases his thinking in a masculine direction, away from a more feminine orientation toward care and relationships (Smiley 2004, 1602–6); (2) that justice as fairness lacks the ability to criticize women's subordination in the private sphere; and (3) that the veil of ignorance requires that people think away key aspects of their identity (1601). Smiley does not find the first criticism to stand up, but neither does she deem Okin's attempt to associate the OP with care and empathy to be successful (1607–8). With regard to the second criticism, Smiley finds that the OP contains the potential to challenge patriarchy in public and private life (1600; cf. 1608–13). The third criticism, which is especially salient for members of oppressed or subordinate racial communities (1613; cf. 1613–19), can be addressed via a reading of Rawls's thought that is, like Laden's, derived from Rawls's increasing tendency to present justice as fairness as a theory for democratic societies. This tendency needs to be strengthened if Rawls's work is to fulfill its feminist and egalitarian potential by emphasizing what it means to belong to a democratic community, with its requirement not just of moral equality but also of political equality and nondomination (1600–1601). What is needed is "a definition of democratic identity . . . that challenge[s] patriarchy and other hierarchical structures of domination" in both public and private spheres (1626; cf. 1620–22, 1624–26). This would, conversely, require that Rawls downplay some other aspects of his thought, such as his view of toleration and the limits of state action, and the role of rational choice (1627).

Susan Moller Okin's "Justice and Gender: An Unfinished Debate" surveys Okin's own, and some other, feminist engagements with Rawls since TJ.[9] Okin also registers her reactions to the later developments in Rawls's thinking about women and the family. Her reaction, once again, is mixed. While grateful that Rawls eventually responded to his feminist interlocutors and adopted some of their recommendations (Okin 2004, 1565), Okin remains troubled by his unwillingness to apply the principles of justice directly to the family (1538, 1563–64, 1567) and puzzled by his comparisons of families to associations (1563, 1566). She insists that a purely political form of liberalism remains unhelpful for feminist purposes (1538, 1555), and finds unconvincing the distinction between respecting women's political equality while propounding their metaphysical or moral inequality. Her answer to Lloyd, Wijze, and Nussbaum, therefore, is that a doctrine promoting the latter cannot truly respect the former (1559, 1562). In this context, Okin reiterates Kearns's original question about how a family that does not practice justice can transmit a sense of justice to its children (1566–67).

Like many feminist commentaries on Rawls, Karen Green's revisits the debate between him and Okin but maintains that Okin's ideals are more destructive of Rawlsian liberalism than Okin appreciates. Four major problems can be identified in Okin's call for a form of androgyny where gender differences are minimized or eradicated. The first is that sex difference might be more biologically rooted than Okin allows, and the second is that some people might feel attached to their gender identities and not want to relinquish them. Pursuing androgyny would, third, interfere with individual liberty by imposing a particular conception of the good on citizens. The fourth problem is that an ideal of androgyny clashes with multiculturalism, for many minority groups have strong commitments to gender difference. A more liberal approach would support a variety of family forms, which is the position that Rawls ends up adopting. Green's preferred way of creating and implementing just family law is to follow a principle of parity that guarantees equal representation of both sexes in parliament. With parity in place, "one can leave the choice between androgyny and gender difference to be determined by individuals within a framework of law that is informed as much by the voices of women as by those of men" (Green 2006).

Catherine McKeen (2006) encourages feminists to be more solicitous of family privacy and family autonomy, for these goods leave adults free to determine the family's comprehensive doctrine. The values and attitudes circulating within a family need not "conform to the dictates of public debate . . . [or be] open to public scrutiny in the first place." These protections against state interference mean that parents can prevent sexist values and attitudes from penetrating the family. Partiality refers to "particular attachments, relationships and loyalties" and is a good that the family realizes. Partiality could even, in McKeen's view, qualify as a Rawlsian primary good. Protecting it means that relationships do not always have to be "answerable to general social concerns," nor must individuals always be "thinking of public concerns and public goods."

In 2006 Lisa Schwartzman published *Challenging Liberalism: Feminism as Political Critique*, underlining the incompatibility between feminist theory and liberal methodology. An abridged version of Schwartzman's critique of Rawls appears here as chapter 2. Schwartzman evinces skepticism about Okin's use of the OP for feminist purposes, for this departs from Rawls by including gender as a relevant social position and attributes knowledge of gender oppression to those behind the veil. Schwartzman also echoes Trout's complaint that Rawls focuses on economic status to the neglect of other forms of social inequality, and follows Trout in advocating a wider understanding of the concepts of "relevant social position" and "least well off." Schwartzman considers, but finds unsatisfactory, the rejoinder that Rawls is doing ideal theory and so writing about inequalities that would be acceptable in a well-ordered society. She also criticizes Rawls for his too permissive pluralism, for some of the conceptions of the good that he tolerates take for granted the continuation of structures of oppression. With its focus on people as individuals rather than as members of classes or groups, liberalism is incapable of perceiving, let alone criticizing, the social hierarchies that subordinate individuals on the basis of group membership. Feminist theorizing, by contrast, must pay attention to collectivities and be contextual and concrete, rejecting also the abstraction that characterizes liberal methodology.

I puzzled through some of the uncertainties that Okin identified in Rawls's later writings on women, and I contend that Rawls goes a long way toward addressing some major feminist-liberal concerns about justice as fairness (Abbey 2007). Yet in accommodating his feminist interlocutors, Rawls pushes justice as fairness in the direction of a more comprehensive, rather than a strictly political, doctrine. Situating the family in the basic structure makes it impossible to confine concerns with freedom, equality, independence, and rights to the strictly political realm. In bringing family relations within the purview of justice as fairness, Rawls must go beyond ideas implicit in society's public political culture, for, as feminists have long argued, the family has traditionally been relegated to the private sphere. Once the family becomes the subject of justice, at least some of the virtues of private life, which Rawls relegates to comprehensive doctrines, become part of his theory of justice. As he fleshes out what it means for women to be equal, free, and bearers of inalienable rights in all (or most) areas of their lives, the insufficiency of a purely political liberalism becomes manifest. I contend that Rawls advocates a type of autonomy for adult individuals in the domestic, as well as in the political, realm, which once again surpasses the limits of a purely political liberalism.

Just as I was suggesting that more than political liberalism was needed to satisfy feminist concerns, Corey Brettschneider advanced a reconstruction that would render political liberalism much more receptive to them. Brettschneider articulates a "principle of publicly justifiable privacy," according to which the public-private distinction must be formulated with reference to political liberalism's values of free and equal citizenship: "to the degree that privacy exists at all, its boundaries must be determined by and normatively argued for through public reason. . . . Domestic life is not immune from political examination" (Brettschneider 2007, 24). Political liberalism's values apply as much to the domestic as to the public sphere and should be used to scrutinize citizens' comprehensive doctrines. When such doctrines clash with those values, the doctrines should be amended. "When public reason demands it, individual moral identity must be completely reformulated" (26; cf. 23, 25). Political liberalism thus has considerable power to transform citizens' outlooks, indicating that Brettschneider offers a much less passive, permissive reading of political liberalism than do interpreters like Lloyd, Wijze, and Nussbaum. Yet Brettschneider insists that political liberalism remains distinct from comprehensive liberalism, for while political liberalism's values constrain and limit family dynamics, they do not dictate how a family should behave or what it should prize or pursue in all areas. This leaves room for a diversity of comprehensive doctrines to address these issues in different ways (27–28). In stark contrast to McKeen's defense of partiality, Brettschneider sees it as a strength of his reading that "the moral identity of individuals as citizens is given priority over their private identities. This often means that our public commitments are more fundamental than private ones" (26).

Like Brettschneider, Baehr (2008) explores the feminist potential of political liberalism and thinks through what it would mean for feminism to become a public political philosophy.[10] While feminism can be classified as a family of partially comprehensive doctrines (195), political liberalism's ambition to be hospitable to the fact of pluralism serves feminism well, for feminism is itself an internally plural school of thought. A robust feminist approach to politics would attract an overlapping consensus from as large a number of women and feminist positions as possible. So Baehr sets out to sketch a form of feminism that would draw its conclusions and recommendations for public policy from the values implicit in the public political culture of a democratic society. As a public political doctrine, feminism would argue for or against certain social practices on the basis of whether they harmed women's liberty and equality as citizens, or how they affected women's physical security or chances of equal opportunity. Baehr indicates what such an approach could justify as well as what its limitations would be. In the concluding chapter of the current volume, she develops her vision of feminism as a public political philosophy, looking closely at liberal feminism through this lens. If liberal feminism can be portrayed as a purely political doctrine, then it can be affirmed by those holding different comprehensive doctrines, such as Jewish feminism, ecofeminism, or socialist feminism.

Continuing the debate about the value for feminists of Rawls's turn to political liberalism, Christie Hartley and Lori Watson contend that political liberalism's commitment to equal citizenship makes it highly supportive of feminism. They distinguish their defense of political liberalism's feminist potential from that offered by Lloyd, Baehr, and Nussbaum by the attention they give to reciprocity. The criterion of reciprocity limits what counts as a reasonable comprehensive doctrine to views that eschew any form of women's subordination to men. They go further than previous feminist defenders of political liberalism by claiming that "political liberalism is a feminist liberalism" (Hartley and Watson 2010, 6). They do not discuss, however, a key issue that arises from attempts like theirs and Brettschneider's to recover a feminist position from political liberalism, namely, the limitations that this necessarily places on pluralism, which was, as noted above, the impetus to its creation in the first place.

The Current Volume

In addition to chapters 1 and 2, by Laden and Schwartzman, respectively, which contain previously published material, this volume offers six new essays that further the debate about the resources that Rawlsian theory provides for feminist thought. Along with chapters 1 and 2, the volume's concluding chapter, by Baehr, which grows out of her 2008 article, has been described above. In chapter 3, Elizabeth Brake mounts an argument about the latent feminist possibilities in Rawls's thought that takes a shape similar to her position on neutrality, described above (Brake 2004). In this volume, Brake explores the place of self-respect in Rawls's thinking. Rawls deems self-respect a primary good, but, in Brake's estimation, having awarded it this cardinal status, he never develops its implications. Brake effectively echoes Trout's (1994) charge that Rawls fails to give one of his own claims the theoretical attention it deserves. Brake notes the ambiguity in Rawls's treatment of this concept and advances a definition of self-respect that holds promise for thinking about gender inequality and the family. She develops the important implications of self-respect for the legal recognition of family forms and for the question of parents' rights to infuse children with their beliefs.

In stark contrast to those who find potential for feminist causes in Rawls's thought, Clare Chambers identifies deep-seated difficulties that emanate from his belief that the basic structure is the appropriate subject of justice. Like many feminist commentators, Chambers revisits the Rawls-Okin debate and considers Okin's suggestion that principles of justice should apply to the family. She examines the rejoinder, first advanced by Lloyd, that the principles apply not to single institutions but to the interactions of the basic structure's components. Chambers replies that what the family's relationship to justice illustrates is that how the principles apply to an institution does not in fact depend on that institution's location in the basic structure. Chapter 4 of the current volume thus interrogates Rawls's wider position that there is something distinctive about the application of justice to the basic structure and finds it unconvincing.

In chapter 5, Nancy Hirschmann considers how Rawlsian theory responds to the claims and needs of disabled citizens. Surveying the debate about this issue, she agrees with feminist critics like Kittay about Rawls's inability to address the crucial issue of dependency and care in human life or politics. But Hirschmann reframes the issue of disability as a question of freedom rather than justice. Examining what Rawls's theory has to offer disabled citizens in the way of freedom reveals the Rawlsian self to be not only masculinist but also able-bodied. She contends that Rawls's work is underpinned by a tacit but influential notion of embodied individualism that makes assumptions about what sort of bodies are able to be free. Rawls's negative conception of freedom, moreover, is ill equipped to respond to claims by the disabled for greater inclusion in social life, and his neglect of the social construction of desires and identities is explained by his general indifference to social relationships. Hirschmann defends the social model of disability and its implications for enhancing freedom by contrasting it with the dominant medical model.

Nussbaum is among the thinkers whom Hirschmann discusses who try to correct for deficiencies in the Rawlsian approach by incorporating disability issues into their thinking about justice. Heavily influenced by Kittay's critique of Rawls, Nussbaum strives to find a way to incorporate human dependency in her capabilities approach to justice. Nussbaum's parallel attempt to show how thinking about justice can be extended globally features prominently in chapter 6, in which Eileen Hunt Botting addresses Rawls and international justice. Hunt Botting examines Rawls's *Law of Peoples* and considers Nussbaum's critique of his position. She suggests that both Rawls's and Nussbaum's theories of justice offer useful resources for advancing feminist values in international and transnational relations, albeit in different situations. Nussbaum's capabilities approach is helpful for defending women's human rights in situations of minimal cultural or religious conflict, whereas Rawls's approach is more useful for women's human rights in situations of significant cultural or religious conflict. Rawls also has the advantage of wrestling with the hard case of using military force to defend women's human rights.

Along with Nussbaum, another feminist thinker to have been powerfully influenced by Rawls in a number of key respects is Jean Hampton. In chapter 7, Janice Richardson examines Hampton's work, comparing and contrasting Rawls and Hampton as contractarian thinkers. Richardson applauds Hampton for applying social contract thinking to the justice of everyday relationships, for this brings this tradition into line with much other feminist thought that is concerned with the implications of everyday beliefs. Richardson aligns Hampton's interest in self-worth with Rawls's references to self-respect, and agrees with Brake about the importance of this concept for feminist thought.

Notes

1. Given the focus of this volume, and in the interest of manageability, it reviews only those writings dedicated to Rawlsian theory. Wider works that are also relevant, such as feminist discussions of the social contract tradition or of liberalism more generally, are not surveyed.
2. See also Kearns 1983 (37) on this omission. In 1975 Rawls added gender to the list of informational restrictions, and he has retained this. See, for example, *PL*, 25.
3. Okin cites and recommends Kearns's work (*JGF*, 196n20).
4. Although her article was published after Rawls's turn to political liberalism, Trout does not engage those developments, which explains the temporary break in chronology.
5. Okin does allow that this element of Rawls's theory could be supposed to simply carry over, unchanged, from *TJ* to *PL*.
6. Because Lloyd republished this material, virtually verbatim, in 1995, citations are given for both articles.
7. Okin had raised this point herself and concluded, much as Lloyd does, that "this question would undoubtedly be helped by some good research" (1994, 38n32).
8. Munoz-Dardé cites English 1977 (341n4) and uses her phrase about the family being opaque to claims of justice (340, 341). She agrees that making contractors heads of families compromises both Rawls's commitment to individualism and his rejection of utilitarian calculations of aggregate benefit. It imports ties of sentiment into the contractors' motivations that are hard for him to justify theoretically. Rawls is also unclear about the extent to which members of one generation care about the welfare of their offspring (Munoz-Dardé 1998, 335, 340–43, 348).
9. Because much of this material is reprised in Okin 2005, I do not provide an independent discussion of that article.
10. Her Rawlsian project suggests a reassessment of her earlier conclusion that Habermas was more useful for feminists than Rawls was (Baehr 1996, discussed above).

1

Radical Liberals, Reasonable Feminists

Reason, Power, and Objectivity in MacKinnon and Rawls

Anthony Simon Laden

As the introduction to this volume shows, John Rawls's turn, if that is what it was, to political liberalism from the comprehensive liberalism of *TJ* was not generally greeted as a welcome turn of events by many feminists.

> This essay has had a very long history, and there are thus many people to thank. I am grateful to John Rawls, Christine Korsgaard, Amartya Sen, Fred Neuhouser, David Peritz, Talbot Brewer, James Tully, Sue Dwyer, Tamar Szabo Gendler, Martha Nussbaum, Dan Brudney, Catharine MacKinnon, Sandra Bartky, Peter Hylton, David Hilbert, David Owen, and the referees for the *Journal of Political Philosophy* for their comments on earlier drafts, and to audiences at the Institute for the Humanities at the University of Illinois at Chicago, the Law and Philosophy Workshop at the University of Chicago, and the philosophy departments at Harvard, DePaul, and the University of Illinois at Urbana-Champaign, and a panel on Rawls and feminism at the Central Division meeting of the APA in 2005, where I presented versions of this material. I was able to complete significant amounts of work on this essay while being supported by the Institute for the Humanities at the University of Illinois at Chicago.

According to the standard interpretation, Rawls, buffeted by the criticisms of communitarians and conservative religious attacks on liberalism, retreated from the bold universal egalitarian theory of his early work to a more limited, cautious, and conciliatory approach in the later work. Among the losers, claim many feminists, were women (Okin 1994; Nussbaum 2003).

What is odd about this interpretation of events is that Rawls's development of political liberalism coincided not only with the increased presence of feminism, both in the academy more generally and in philosophy in particular, but also with an increasing concern on Rawls's part with feminism and feminist critics of his own work. Moreover, whereas his treatment of the major communitarian criticisms of his work is dismissive at best, his reaction to feminist critics shows that he took them much more seriously. If you compare his footnote references to the work of Sandel (PL, 27, 388n21) with those to the work of Okin, Nussbaum, and other feminists (IPRR 156–57n58), it certainly looks as though he was much more concerned with developing a feminist-friendly liberalism than a communitarian-friendly one.

In this chapter, I argue that Rawls succeeded, at least in one, perhaps surprising, respect. I argue that political liberalism is not blind to the oppression of women in the way that Catharine MacKinnon claims that other forms of liberalism are. MacKinnon criticizes liberalism for, among other things, its reliance on a norm of objectivity. I start by unpacking this claim with the aim of showing its merit, both as a philosophical criticism of some forms of liberal political philosophy and as a criticism with political ramifications. Political liberalism, however, does not rely on the norm of objectivity that MacKinnon's criticism targets, and thus it need not be blind to the oppression of women. Moreover, as I argue, its account of political justification provides a norm of objectivity that feminists could comfortably adopt.

To say that political liberalism is not blind to the oppression of women is not to say that it has no flaws to which feminists might object. Other chapters in this volume raise concerns about some of these flaws. It is rather to say that it is an approach with which feminists can enter into productive conversation, rather than a theory that they are doomed always to be talking past. To see the scope of my claim, it helps to distinguish two forms of faulty vision. A theory is shortsighted if it fails to attend to a form of oppression. Shortsighted theories can be corrected by pointing out the form of oppression missed. A theory is blind, however, if it is not theoretically equipped to respond to such a point, if there is nothing within its theoretical machinery that allows it to see the oppression it is shown as oppression. Thus a theory developed to address injustice that takes the form of inequality in income and wealth may have nothing to say about workplace democracy or the role of disability or dependency in creating injustice. If these result from mere shortsightedness, then the theoretical framework can be used to address these issues as well. Nothing in the theory makes it hard to conceive of a lack of workplace democracy or the failure to accommodate forms of disability or adjust for dependency as forms of injustice. If, in contrast, the theory is blind, then something about its conceptual framework occludes the very possibility of conceiving of these disadvantages as forms of injustice. These failings often go hand in hand, but they are independent. A theory can be relatively broad of vision and yet have blind spots, and a theory can be shortsighted but not blind.

MacKinnon charges liberal theory and jurisprudence with being both shortsighted and blind. Her criticism of liberal objectivity concerns its blindness, however, and it is this criticism that concerns me here. In claiming that political liberalism is not blind to the oppression of women that MacKinnon's work highlights, I do not rule out the possibility that it is nevertheless shortsighted with regard to that very oppression. The conclusion at which I aim does not deny that there is much work to be done in both seeing and remedying injustice to women. It merely claims that such work might profitably be done in conversation with Rawls's political liberalism.

Liberalism, Objectivity, and the Male Point of View

MacKinnon's philosophical criticism of liberalism starts from a political insight. Liberalism treats sex inequality as discrimination rather than oppression.[1] It is thus blind to forms of oppression that do not arise from discrimination. According to MacKinnon, liberalism adopts a "difference approach" to sex equality. The difference approach holds that there is inequality when people who are relevantly similar are nevertheless treated differently. Liberals who challenge discrimination against women do so by holding that men and women are similar in all relevant respects, and so any differential treatment counts as inequality (MacKinnon 1987, 32; MacKinnon 2005). The liberal's feminism, then, rests on the claim that gender differences are thus "morally irrelevant."

MacKinnon, by contrast, claims that sex inequality is a matter of the systematic oppression of women by men (1989, 3). Thus establishing real sex equality involves more than working out the "relevant" differences between men and women and rooting out unwarranted discrimination. It involves, rather, abolishing the relations of domination and exploitation of sexuality that both oppress and define women. MacKinnon's scathing attacks on liberalism rely on the thought that rooting out discrimination often does nothing about the underlying exploitation, and more often than not obscures the real problem. The failure to focus on oppression rather than discrimination might at first appear to be a claim that liberalism is shortsighted, merely failing to take account of an important site of unequal distribution, namely, the distribution of power. MacKinnon, however, thinks that the problem goes deeper[2] and argues that liberalism's failure to recognize and address the oppression of women is due to its blindness. She traces this blindness to its reliance on norms of objectivity (1987, 42). Very roughly, the problem is that liberals conceive of political justification in a manner that leads them to ask the wrong questions when investigating sex inequality. Because their theory leads them to ask the wrong questions, they are bound to overlook forms of sex inequality not revealed by those questions. And the inequality they overlook is that which does the most harm to women.

To see why this amounts to blindness and not mere shortsightedness, we have to understand what moves liberalism to ask the questions it does. The answer, according to MacKinnon, is its theory of political justification, a theory I call rationality as objectivity. Rationality as objectivity holds that it is rational, and thus justifiable, to rely on social categories only if they are properly rooted in objective facts. Facts are objective in this sense when they are prepolitical in the sense that they are regarded by the theory as holding independently of political or social arrangements rather than being a product of them. A political theory might rely on rationality as objectivity by claiming that political categories must track natural facts. Natural facts are taken to hold prepolitically, to lie in the background, and so to be objective. Of course, most contemporary liberals reject appeals to what is natural. They nevertheless rely on a similar structure of justification that replaces the "natural" with the "morally relevant," or, in the words of Rawls's earlier work, what is not "arbitrary from the moral point of view" (*TJ*, 15).[3] This leads such theorists to be particularly concerned with distinguishing between the differences among people that justify differential treatment (often labeled "choices") and those that do not (often labeled "circumstances") (e.g., Barry 2000).[4]

Theories that rely on rationality as objectivity rest on an assumption about what they take to be the relevant set of background facts. Not only are such facts taken to obtain prepolitically, but so is their status as relevant or morally arbitrary. As a result, the relevance of such facts holds independently of any particular political or social structure, and their importance is not to be explained by reference to any such structure. Thus, for instance, if men are naturally different from women, this is true no matter how radically we revise society. And if gender is arbitrary from a moral point of view, then it is as irrelevant in justifying political principles or policies in a rabidly sexist society as in one marked by gender justice.

The importance of there being a set of fixed considerations upon which the justification of political principles and policies must proceed makes justice a matter of mirroring. That is, just distributions turn out to be ones that properly reflect the background facts declared relevant. If, for instance, we want to know if a social category like gender is one that our theory needs to pay attention to, we need to ask about the significance of the correlating background category—in this case, sex. But the significance of background facts is presumed to rely only on information contained within the realm of those facts. So, to continue the example, if we want to know if sex is a significant background category, one with relevance to political justification, we need to figure out whether any background facts, including the sex differences themselves, explain the significance of sex characteristics. We might, for instance, point to the different and complementary roles the two sexes play in reproduction as an explanation for why sexual differences are intrinsically important. Similarly, we might point, as many liberals do, to the lack of differences between men and women in rationality, autonomy, or other essentially human attributes to determine that sex differences are morally irrelevant, and thus not significant background facts. Rationality as objectivity claims that it is rational to base political decisions on social categories only when they rest on significant background (objective) facts. If a social category is to provide the basis for a politically valid reason, it must be morally relevant, and thus it must correlate with a significant background fact. The liberal reliance on rationality as objectivity thus explains why liberals think about sex equality in terms of discrimination. Their theory of political justification tells them to investigate suspect political decisions by evaluating the relevant background facts. Faced with the differential treatment of men and women, such a theory tells us to ask whether there is any background fact that could support such treatment. In other words, it leads us to focus on questions of differential treatment and whether that treatment properly mirrors the right background facts, which is to say, whether it is discriminatory.

MacKinnon's criticism of rationality as objectivity starts from her claim that, at least for gender, the basic assumption behind rationality as objectivity is mistaken: for gender, it is the social world that is primary. In other words, gender is "socially constructed." When a category is socially constructed, its significance is not traceable to, or justifiable by, any set of correlating background facts. Social power constructs such categories directly, by organizing them and carving up the social world using them, and thus marking them as significant (MacKinnon 1989, 113). Since the significance of a socially constructed category does not arise from its connection to certain background facts, rationality as objectivity has us barking up the wrong tree. Proper evaluation of socially constructed categories requires investigating their content and the social forces that are necessary to reproduce them. Gender, according to MacKinnon, is a hierarchical relation reproduced in large part through violence against women. These facts about gender, rather than its lack of correlation with natural or morally relevant facts, account for its injustice.

Thus, with regard to gender and other socially constructed categories, rationality as objectivity turns our attention to the wrong phenomena and leads us to ask the wrong questions. Faced with a socially constructed social distinction, rationality as objectivity tells us to ask, "Is this social distinction based on a relevant objective difference?" MacKinnon's point is that this cannot possibly tell us what we need to know. Instead, we should ask, as MacKinnon's theory suggests, "Is this social distinction unjust?" Answering that question requires looking at the social distinction itself, as it is embedded and reproduced in social and political practices and institutions. Note here that rationality as objectivity leads not to mere shortsightedness but to true blindness. Because it leads to a concept of justice as mirroring, as a matter of who has what, it has no way of conceiving of the mere fact of domination and subordination as unjust (see also Young 1990).

Rationality as objectivity thus blinds us to certain forms of oppression because it leads us to ask the wrong kinds of questions, questions that deflect our attention from the injustices built into certain social categories. Furthermore, such blindness is not innocent or arbitrary. It turns out to be deepest when oppression is the most severe. In these cases, the congealing effects of social power are the least contested, and thus the correspondence between the background facts and the social ones is tightest. A society in which women are not permitted to engage in certain activities is also one in which they are not trained for such activities or taught to conceive of them as among their options. Their lack of training or ambition, however, then serves as a justification for their exclusion that makes no reference to the morally irrelevant fact of gender. Oppression can thus be self-justifying, because to the extent that it is effective, there appears to be the strongest of rational bases for discriminatory policy. It is, for instance, possible to disadvantage women systematically with a set of policies that never mentions sex. Such policies only need to rely on the "morally relevant" characteristics that a patriarchal society has implicitly designated as masculine or feminine (MacKinnon 1989, 162).

The problem with rationality as objectivity is not that it misdescribes what it regards as the background facts. Rather, its mistake is to give them a role in justification that hides the source of the significance of their social correlates. Adopting the norm of rationality as objectivity leads us to make what Sally Haslanger describes as a modal mistake (1993, 103). We see the significance of the background facts in question as arising from natural or inevitable features of those differences themselves, rather than as a consequence of the organization of social power.

The problem to which MacKinnon's analysis calls our attention is thus different from a power-related problem about which liberals have been more concerned. Liberals who rely on rationality as objectivity do worry about whether their description of the set of background facts is accurate, and thus whether their particular theories are not as objective as they originally thought. And they have often rightly explained these shortcomings by reference to inequalities in social power. That is, they can and do recognize that the organization of social power has left previous liberal theories misdescribing what the background facts really are, and thus what sorts of differential treatment might be justifiable, and they are generally open to the possibility that they are making similar mistakes. These, however, are problems of shortsightedness. They are, in theory, correctable by pointing out the distortion that social power establishes in *our view* of what we take to be the appropriate background.[5] So, for instance, while we used to think that women were unfit for certain activities on the basis of their biology, we now realize that this was due to a lack of training and encouragement. This involves a shift in the *content* of the background facts without abandoning the general form of justification.

MacKinnon's criticism goes deeper, however. If gender is entirely socially constructed, then there is no set of prepolitical background facts whose correlation with facts about gender could justify our reliance on them. The problem, as MacKinnon sees it, is not that liberals have gotten the content of the background facts wrong, but that they have relied on them for political justification at all. What is needed, then, if liberalism is to open its eyes to the systematic oppression of women, is not further refinement of the category of "moral irrelevance," or discussions of what, precisely, is "arbitrary from a moral point of view." We need, instead, a wholly different conception of political justification. It turns out that *PL* offers us one.[6]

Reasonableness in Political Liberalism

Although he may not always have fully realized the implications for the argument presented here, Rawls never adopted rationality as objectivity or a mirroring conception of justice. For Rawls, justification is not a matter of relating actions or principles to static background facts but is always an intersubjective practice. Justification is always, for him, justification *to* another (*TJ*, 580).[7] Since this feature of Rawls's work is most clear in the argument of *PL*, I focus my remarks on this later work and its conception of legitimacy.

Political liberalism is designed to respond to what Rawls calls "the fact of reasonable pluralism" (*PL*, 24n27, 36–37). The consequence of reasonable pluralism is that there can be no nonoppressive agreement, even among reasonable citizens, about comprehensive philosophical or moral principles. Thus, if political principles are to avoid being oppressive, they cannot be justified in terms of a particular comprehensive doctrine. Political principles that require particular comprehensive doctrines for their support rest on reasons that not all reasonable citizens accept. When a state acts on such principles, it thus violates what Rawls describes as those citizens' "political autonomy" (xliv, 77–81).[8] For them, the justification of the use of state power is arbitrary and thus alien. Legitimate political principles ought not to undermine any reasonable citizen's political autonomy. They must thus rest on reasons that each reasonable citizen can regard as authoritative, given what Rawls calls his or her "comprehensive doctrine" (xviii–xix, 13).[9] Such principles are then reasonable.

What determines whether all reasonable citizens can endorse a set of principles is whether they emerge from a joint activity of what I call reasonable political deliberation. That is, the reasons that can serve to justify political principles are not those that bear the right sort of relationship to some prepolitical set of background facts, but those that emerge from a political process of reasonable deliberation. Rawls thus rejects rationality as objectivity. Instead, he suggests that for political purposes we must regard the authority of reasons as depending on the person to whom they are offered, and in particular on his or her general philosophical, moral, or religious outlook. What exactly does this mean? I think it helps to think of reasons—in particular, reasons for action—as authoritative claims that we make on one another and on ourselves. For such a claim to have authority in the eyes of the person on whom it is urged, it must rest on something that she regards as a ground of authority. Rawls describes this ground of authority as a conception of the good.

Imagine that I make a claim on you that I convincingly demonstrate is authorized by the doctrine of a particular church. If you are a member of that church, you will probably take my claim as having sufficient authority. But if you are not religious in this way, my claim has authority for you only if it can rest on other grounds. So far, this is all rather commonplace. What is significant about Rawls's view of reasons is that, ultimately, nothing deeper, more objective, needs to or could ground the authority of reasons in political deliberation. If someone's comprehensive conception of the good rejects all possible grounds of authority for an offered reason, then if the state nevertheless rests the justification of a political principle on that reason, it violates her political autonomy. Suppose that I oppose the legal availability of abortion, and that the only reason I can give for my position rests on a religious account of the sanctity of human life from the moment of conception. Were the state to adopt my position on the basis of this reason, it would violate the political autonomy of people who have a different or no religious belief and thus reject this premise. The state's action would be unjust because unreasonable.

I want to expand this point by suggesting a shift in language. Rawls talks of reasons' being grounded in "conceptions of the good." The phrase "conception of the good" is often read as referring to a set of beliefs, such as a religious or philosophical doctrine, but I think that Rawls conceives of it in rather broader terms (*PL*, 108, 303). Christine Korsgaard talks of reasons' being grounded in what she calls "practical identities" (1996, 101).[10] "Practical identity" captures that breadth. My practical identity consists of those things about myself that I value and regard as grounding my obligations. It can thus include my religious or philosophical affiliation, but also my gender or ethnicity or race, where these are understood not as objective facts about me but as social ones.

This shift in terminology brings to the surface a particular form of unreasonableness that is relevant to feminist concerns. We treat someone reasonably when we take seriously their rejection of claims we make on them, when we acknowledge that what we take as authoritative grounds for some claim may not be authoritative for them, when we accept that "no" means "no." By thinking of the authority of reasons in terms of practical identities rather than conceptions of the good, we make it easier to recognize the variety of ways in which people fail to be reasonable. It is unlikely that I will assume that your conception of the good includes the tenets of a particular religious faith if you have manifested no sign of holding such beliefs. Faced with someone who shows no sign of belief in the tenets of Catholicism, say, we say that he is not a Catholic, not that he is a failure as a Catholic. It is, however, quite common for people to assume that someone's practical identity includes a particular gender role on the basis of biological facts about him, rather than because of any manifestation that the person endorses that role. Faced with someone who is biologically female and yet clearly rejects many or even all of the traditional roles that go along with being gendered female, we are much more likely to see her as somehow failing to live up to (or perhaps challenging or resisting) an identity that is automatically, ineluctably hers. Deliberating with someone under such an assumption leads us to assume that certain claims have authority for her when they do not. It also prevents us from taking her rejection of those claims seriously. We thereby treat her unreasonably.

The importance of noticing this form of unreasonableness becomes clear when we turn to Rawls's account of reasonable political deliberation. Political liberalism holds that political deliberation must satisfy two criteria if it is to be reasonable. The first, which is at the forefront of Rawls's work, requires that in political deliberation citizens offer as justification to one another only those reasons that they think in good faith their fellow citizens could regard as authoritative. Rawls calls these reasons "public reasons" (PL, 212–54; IPRR). In political deliberation, we address one another as citizens. We offer reasons to one another that we sincerely believe that all our fellow reasonable citizens can share despite their diversity. It is thus unreasonable to offer reasons that cannot find support in some of our fellow citizens' varied practical identities.[11] But it is also unreasonable to offer reasons whose support rests on aspects of a practical identity that are incompatible with citizenship, even if it is an identity that some of our fellow citizens endorse.

The second criterion is less explicit in Rawls's work. Deliberation is reasonable only if the rejection of a claim is taken seriously, if it alters the future course of the deliberation. When a citizen rejects a reason, her rejection must be taken seriously by her fellow citizens. It cannot merely be ignored or explained away as a result of her own failures and pathologies.[12] This places constraints on social relations more generally. Although this criterion is not at the forefront of Rawls's discussion of public reason, it is at work in his theory, and it emerges most clearly in his requirement that reasonable deliberation satisfy the criterion of reciprocity, which he describes as follows: "The criterion of reciprocity requires that when those terms are proposed as the most reasonable terms of fair cooperation, those proposing them must also think it at least reasonable for others to accept them, as free and equal citizens, and not as dominated or manipulated, or under the pressure of an inferior political or social position" (IPRR, 137).

I pointed out above one way in which both women and men can be unreasonable with respect to women: when they assume that a woman has a certain practical identity as a result of her biology. By focusing on this second criterion of reasonable deliberation, we can see how the adoption of a norm of rationality as objectivity both hides and exacerbates such unreasonableness. A man makes an assumption that a woman rejects. Since his assumption is grounded in objective facts, rationality as objectivity says that the assumption is rational. It thereby also concludes that she is unreasonable. From this perspective, the man's assumption and further behavior seem justified. If we look at the same situation from the point of view of Rawls's theory of reasonable deliberation, however, things look different. The man has raised a claim and the woman has rejected it. He fails to take her rejection seriously. So he is unreasonable, not primarily for having made the assumption but for failing to take seriously an argument that would invalidate it. Thus whatever follows from his assumption cannot ground legitimate political principles within political liberalism.

Feminists routinely point to the ways in which gender is politically reinforced so that gendered traits are made morally relevant. They then conclude that justice demands the unmaking of this aspect of gender. But these demands are often dismissed by liberals who rely on rationality as objectivity. The liberals point to the same phenomena but take them as evidence of the moral relevance of the trait in question. The liberals thus take these traits to support the justice of the current political practice. Since the feminist is not asking the questions that rationality as objectivity instructs the liberal to ask, she is seen as not raising appropriate concerns about the justice of current arrangements. But according to the theory of reasonable deliberation on which Rawls relies, it is the liberal's attitude, and its supporting norm of rationality as objectivity, that undermines the possibility of reasonable deliberation: it is those dismissive of the feminist complaint, whether in their guise as men or as liberals, who are unreasonable.

Reasonable Feminism Meets Radical Liberalism

We can now see how feminist claims about the political importance of the social oppression of women get a hearing within political liberalism. Within political deliberation it is always a legitimate move to reject any reason whose authority rests on a practical identity that is incompatible with being a citizen. Freedom and equality are necessary features of citizenship.[13] Freedom and equality are inconsistent with social hierarchies maintained by violence. Thus unjust social categories, because of their injustice, are unsuitable grounds for reasons urged in political deliberation. It is reasonable to reject political principles that rely on such categories. Since such rejection need not rest on claims about the objective world, we can reject reasons that are based on objectively grounded but nevertheless unjust social categories. Note that political liberalism leads us to ask what MacKinnon claims is the right sort of question.

As implied above, it is a necessary condition of deliberation's being reasonable that such moves must not only be formally possible but must be taken seriously within the deliberation. This condition requires that the criterion of reciprocity be met in the wider society, and that political deliberation be sensitive to the kind of eye-opening philosophical work that challenges the failures of citizens to acknowledge and address certain forms of oppression. Feminists working within political liberalism can thus point to male-dominated society and the conception of political justification at work in current reigning political theories as barriers to political legitimacy. Thus political liberalism provides avenues by which feminists can champion not only the transformation of the social category of gender but also the social and theoretical dynamics that stand in the way of such a challenge.

The first of these strategies describes what might be thought of as MacKinnon's substantive arguments for various policies and legal reforms. Take her argument for the regulation of pornography. It rests on the claim that pornography helps to construct the social categories of gender in oppressive ways. Pornography thus contributes to a social world in which women are prevented from truly being citizens. In general, seeing her substantive positions in this light explains her insistence that the policies and practices that her feminism rejects are violations of women's civil rights: rights that women require in order to be citizens. Within political liberalism, such an argument is reasonable. Similarly, the second strategy captures the point of MacKinnon's more methodological arguments, such as the criticism of rationality as objectivity. This argument points to the reaction to the first sort of argument as grounds for thinking that the social conditions under which reasonable deliberation is possible do not obtain. Feminists make substantive arguments on the basis of the way certain policies and institutions construct gendered roles as incompatible with citizenship. These arguments are misconstrued, regarded as evidence of the unreasonableness of feminism, or merely ignored. But the ability to have such arguments taken seriously is itself a condition of political deliberation's being reasonable. So the reception of these substantive arguments is a sign that political deliberation is not reasonable. Insofar as the unreasonableness of political deliberation is, in part, a result of its participants' implicit reliance on rationality as objectivity, the criticism of this norm is an attempt to make reasonable political deliberation possible. To that extent, the feminist's methodological argument is also reasonable.

Political liberalism is thus not blind to the oppression of women. It regards the radical demands of MacKinnon's feminism as reasonable. This does not mean that it endorses all of the substantive proposals that MacKinnon endorses or gives them the priority that she gives them. What a politically liberal society winds up doing once it takes arguments like MacKinnon's seriously depends on a host of other details. My claim here is that within the framework of political liberalism, the full force of MacKinnon's arguments on behalf of women's equality can be felt. I am not here making any claims about what might happen once that force is felt.

Feminist Objectivity

MacKinnon describes the project of feminism as "to uncover and claim as valid the experience of women, the major content of which is the devalidation of women's experience" (1989, 116). This suggests that political liberalism shares its conception of objectivity with MacKinnon's feminism (Cornell 1995, 178–83). The norm of objectivity on which political liberalism relies does not require that we take up the view from nowhere, but rather that we adopt the perspective of the reasonable citizen: the person committed to legitimating political principles via reasonable political deliberation. To quote Rawls:

> To say that a political conviction is objective is to say that there are reasons, specified by a reasonable and mutually recognizable political conception . . . , sufficient to convince all reasonable persons that it

is reasonable. Whether such an order of reasons actually obtains, and whether such claims are in general reasonable, can only be shown by the overall success over time of the shared practice of practical reasoning by those who are reasonable and rational and allow for the burdens of judgment. Granted this success, there is no defect in reasons of right and justice that needs to be made good by connecting them with a causal process. (PL, 119)

Call this view "objectivity as reasonableness." Unlike rationality as objectivity, it does not look to background facts to determine objectivity. Rather, the objectivity of political principles depends on whether they would find support in the course of reasonable political deliberation. Political principles that support, maintain, or enforce unjust gender roles are incompatible with at least the citizenship of women. This incompatibility is what undermines their reasonableness and thus their objectivity.

Adopting this view requires neither treating all invocation of gender in political deliberation as inappropriate nor determining which invocations are relevant on the basis of whether or not they have a rational basis in a set of background facts. Rather, it requires asking two questions: first, whether the social categories that define gender are compatible with citizenship, and second, whether their rejection for failing to be so compatible is something that could matter. Answering the first question requires that we look at the details of women's lives as they are lived under a social system that subordinates them. It requires that we see those details not as confirming disabling stereotypes but as pointing toward injustices that need to be addressed. Finally, it requires using, where necessary, the powers of the state to address these injustices.

Answering the second question involves working out a political theory in which asking the first question can be seen as legitimate, as reasonable, and using that perspective to look at people's claims to justice. I have argued that political liberalism is such a theory. Such a theory is also what comes out of adopting what MacKinnon calls feminist method, a method she describes as also avoiding the trap of the norm of rationality as objectivity: "The reality of women's oppression is, finally, neither demonstrable nor refutable empirically. Until this problem is confronted on the level of method, criticism of what exists can be undercut by pointing to the reality to be criticized. . . . If this analysis is correct, to be realistic about sexuality socially is to see it from the male point of view, and to be feminist is to do so with a critical awareness that that is what one is doing" (1989, 124). In other words, as long as we take up the objective stance characterized by rationality as objectivity, we only see the reality of women's lives as proof that the world could not be any other way. The feminist alternative is to see exactly the same facts, but to see them as signs of injustice. Political liberalism can also see these facts with this critical awareness. In doing so, it takes not the standpoint of men but that of citizens. It is from the standpoint of citizens that political liberalism authorizes a rejection of hierarchical social relations, including those based on gender. But it is also from that standpoint that we uphold the norm of objectivity as reasonableness. Such a standpoint is both radical and liberal, both reasonable and feminist.

By bringing Rawls and MacKinnon into dialogue about the question of objectivity, this chapter has shown that there is considerable feminist potential in Rawls's thought, even after his turn to political liberalism. Even though it may be accused of shortsightedness, political liberalism is not blind to the oppression of women in the way that MacKinnon, a leading feminist critic of liberalism, claims other forms of liberalism are. Political liberalism offers an approach to politics and justice with which feminists can and should enter into productive conversation.

Notes

1. MacKinnon is not alone in criticizing liberalism for a focus on discrimination rather than oppression. See, for instance, Young 1990, especially Young's criticism of what she calls the "distributivist paradigm." As will become clear below, I think she misdirects that criticism at Rawls, but I take the general point of the criticism to be important.

2. Young 1990 makes a similar point: the distributivist paradigm leads to a treatment of power as if it were a distributable good, which loses sight of relations of domination.

3. Interestingly for my later argument, such language does not appear in *JFR*, where Rawls defends the two principles on something much more like the grounds, I argue below, that MacKinnon suggests.

4. Barry's reliance on the moral irrelevance standard is clear in his attack on multiculturalism, which relies to a large degree on the claim that, unlike gender and race, one's cultural attachments are morally relevant.

5. I thank Martha Nussbaum and Dan Brudney for pushing me to be clear on this point.

6. Although I focus on *PL* in this chapter, I think that the basic ingredients of the conception of justification I discuss are present throughout Rawls's writings. For an elaboration on that conception of justification, see Laden 2010.

7. Developing political philosophy in light of something like this conception of justifiction has occupied many political philosophers in recent years. For an overview of that work in all its variety, some following Rawls, others situated differently, see Chambers 2010.

8. Rawls generally uses the phrase "the full autonomy of citizens," which he regards as "political, not ethical."

9. Note that this requirement can be met even if the reason in question is not, as it were, derivable from my conception of the good.

10. Although Korsgaard offers a comprehensive moral account, I do not mean to take on board that whole edifice in using the term to describe Rawls's project. Rather, I am suggesting that Rawls

can be read as claiming that for the purposes of political deliberation, we must deliberate as if it were the case that our practical identities served as the ground of the authority of reasons for us.

11. A conception of the good might endorse a political reason only indirectly, and it is a mark of reasonable conceptions of the good that they are willing to do this. Thus, for instance, I might not find certain reasons convincing, but I could nevertheless endorse their results because they emerged from political deliberation that I regard as reasonable.

12. I have discussed the importance of this criterion to a theory of reasonable deliberation in Laden 2000, and have developed more fully the radical consequences of bringing this criterion to the forefront of a political theory in Laden 2001.

13. Rawls discusses the freedom and equality of citizens in *PL* (19, 30). For an argument that grounds the freedom and equality of citizens in their role in reasonable political deliberation, see Laden 2001, 114.

2

Feminism, Method, and Rawlsian Abstraction

Lisa H. Schwartzman

As indicated in the introduction to this volume, Rawls has been enormously influential in contemporary political philosophy, and his ideas and methods continue to shape philosophical debates about justice, liberty, equality, autonomy, and democracy. Given the centrality of Rawlsian concepts to these debates, feminist political theorists must consider how and whether Rawls's work can be employed for feminist ends. Because he moves beyond a narrow focus on individuals and emphasizes the importance of the "basic structure" of society, Rawls's theory of justice—and his later work on political liberalism—may appear promising to feminists seeking to rectify inequalities in society's institutions and structures of power.

In this chapter I raise a number of problems with feminist attempts to use Rawls's work to argue for feminist conclusions. My specific interest is in the question of whether an abstract, ideal theory like Rawls's, where we are instructed to think from the OP, is an effective feminist strategy, and I conclude that it is not. I begin by examining Susan Okin's suggestion that Rawls's theory of justice can be reformulated in ways that hold great potential for feminism. I argue that Okin's feminist reformulation of Rawls depends on two assumptions: (a) that the parties in the OP must think from the perspective of persons living in the actual social positions of our current society, and (b) that the parties know that women are subordinated in current society and recent history, and that they understand the experience of gender oppression. I argue not only that these assumptions are contrary to Rawls's own description of the OP, but also that they are precluded by the sort of idealized theorizing that Rawls advocates. While I think that Okin is correct to suggest that we learn a great deal by thinking about what women would agree to, and by thinking from the perspective of the most subordinated social groups, I find unconvincing her arguments that Rawls provides us with many resources to aid in this project. In the final sections, I raise further problems with assumptions that both Okin and Rawls make about the "pluralism" of views that would exist in a well-ordered society, and I suggest that some of these assumptions take for granted the continuation of structures of oppression. I conclude by suggesting that feminism requires identifying and challenging socially generated hierarchies of power, which Rawls's methodology is ill equipped to do, given its understanding of power, equality, and social position.

Okin's Feminist Reformulation of Rawls

As the introduction to this volume makes clear, Okin acknowledges many problems with PL (Okin 1994). However, as also indicated there, she believed that some of the basic ideas presented in TJ held great promise for feminism: "The original position, with the veil of ignorance hiding from its participants their sex as well as their other particular characteristics, talents, circumstances, and aims, is a powerful concept for challenging the gender structure" (JGF, 108–9). Because persons in the OP know that they could end up as women, they would not agree to women's subordinate status in the family, workplace, or any other sociopolitical institution. Okin contends that "a consistent and wholehearted application of Rawls's liberal principles of justice can lead us to challenge fundamentally the gender system of our society" (JGF, 89). Although Rawls does not mention ignorance of one's sex in his initial description of the veil of ignorance, he adds sex—along with race—to the list of unknown features in his 1975 essay "Fairness to Goodness," which suggests that his theory could be used to challenge sex and race inequalities.

Departing from Rawls's communitarian and feminist critics, Okin argues that his theory does not require an egoistic and individualistic subject who thinks only about impartial and universalist principles. Although Rawls's use of the language of rational choice suggests this reading, Okin claims that "the original position and what happens there are described far better in other terms" (JGF, 101). In particular, the parties would have to act out of "equal concern for others," which would involve paying careful attention to the specific experiences of persons living in the various groups to which one must imagine oneself belonging (Okin 1989b, 246): "the only coherent way in which a party in the original position can think about justice is through empathy with persons of all kinds in all the different positions in society, but especially with the least well-off in various respects. To think as a person in the original position . . . is to think from the point of view of everybody, of every 'concrete other' whom one might turn out to be" (1989b, 248). Okin adds that since the parties in the OP would understand the "general facts about human society," they would know that society is "gender-structured both by custom and still in some respects by law" (JGF, 91), and "that women have been and continue to be the less advantaged sex in a great number of respects" (JGF, 102–3). This knowledge, along with the requirement that they think about justice from the point of view of "everyone," would lead the parties in the OP to consider the possibility that they could be women, and to choose principles that would bring about gender justice.

The Centrality of Economic Categories

Although Okin believes that his theory can be modified to account for gender, Rawls, in his own articulation of the difference principle, refers almost exclusively to income, class, and economic status, paying far less attention to other forms of social inequality. While he claims that his second principle of justice (which includes both the difference principle and the principle of fair equality of opportunity) applies to all social and economic inequalities, it is the class positions defined by income and wealth that are of most concern. Explaining the requirement that those in the OP evaluate the basic structure from the point of view of equality and of representative citizens who occupy "relevant social positions," Rawls contends that each person must consider two perspectives: the position of "equal citizenship" and the perspective "defined by his place in the distribution of income and wealth" (*TJ*, 96). He also emphasizes economic class in his account of the difference principle and in his explanation of how we are to determine which group is the least well off (*JGF*, 97–98). While this may not be easy to establish, the difficulties, as Rawls explains them, concern questions about average versus relative income and wealth, and similar questions that arise within a class-based analysis. This is troubling for feminists, as well as for those concerned about racism, heterosexism, and other forms of oppression, since Rawls seems to focus his discussion of inequality on questions of economic class, thereby rendering other forms of social power invisible. The difficulties that feminists might have in defining the "least well off" involve important questions about how racial, class, and gender oppression should be weighed and balanced; Rawls, however, does not even consider such questions.

TJ does include one passage where Rawls explicitly raises the possibility that social positions other than "the various levels of income and wealth" may sometimes need to be taken into account:

> If, for example, there are unequal basic rights founded on fixed natural characteristics, these inequalities will single out relevant positions. Since these characteristics cannot be changed, the positions they define count as starting places in the basic structure. Distinctions based on sex are of this type, and so are those depending upon race and culture. Thus if, say, men are favored in the assignment of basic rights, this inequality is justified by the difference principle (in the general interpretation) only if it is to the advantage of women and acceptable from their standpoint. . . . Such inequalities multiply relevant positions and complicate the application of the two principles. On the other hand, these inequalities are seldom, if ever, to the advantage of the less favored, and therefore in a just society the smaller number of relevant positions should ordinarily suffice. (*TJ*, 99)

This passage is also one of few in *TJ* in which Rawls explicitly mentions sex as a possible "relevant social position." Within a sentence or two, however, he concludes that since inequalities based on sex and race are seldom to the advantage of women or people of color, these would not *be* relevant social positions in a just society. In a just society, there would be a group of people who would be the "least well off," and they would be the ones whose interests must be taken into account when justifying social and economic inequalities. Yet women, people of color, and others who suffer from various noneconomic inequalities seem not to be included in this group because (presumably) these would not be politically relevant positions *in a well-ordered society*. In contrast, Rawls does assume that some class differences are justified by the difference principle and that the well-ordered society would therefore contain economic classes. These would be justified because they would benefit the least well off class through advantages such as increased incentives and greater productivity.

In response to my argument, one might contend that Rawls's theory precludes sexual and racial inequalities because it is an abstract ideal that prohibits inequalities based on features that are not morally salient. Not only does Rawls guarantee that people have equal basic liberties, but he also writes about the importance of ensuring the "fair value" of these equal liberties, which must be more than merely formal guarantees (*TJ*, 225). Thus one could defend Rawls by arguing that his theory requires women's equality, as well as the equality of various racial groups, and that only certain kinds of class inequalities are permissible (i.e., those benefiting the least well off economic class). The problem with this argument is that it suggests that omitting any discussion of sex and race is a way of ensuring sexual and racial equality. However, insofar as sex and race define socially salient categories in a basic structure, these are important social markers of inequality. Although an abstract ideal can—and should—posit a world in which sex and race no longer categorize people into socially subordinated groups, a theory that makes no (or very little) reference to these social categories effectively suggests that they are unimportant in theorizing about justice and equality.

Ideal Versus Nonideal Theory

A Rawlsian might also argue that because *TJ* deals mainly with ideal theory, that is, a world in which there is "full compliance" with a theory of justice (as opposed to "partial compliance"), Rawls need not address the specific forms of injustice based on factors such as gender and race.[1] Silence on these matters in a discussion of *ideal theory* need not indicate that Rawls views these as insignificant forms of injustice; rather, the appropriate place for a discussion of these injustices would be in the context of nonideal theory, which involves a world in which there is "partial" compliance with the ideal of justice. There are several problems with this line of argument, however. First, the process of devising an ideal is never entirely independent of the context of one's own (nonideal) society. Thus Rawls's understanding of the different social positions present in his ideal theory of justice is necessarily affected by his view of the current social positions in our own society. It would be impossible to come up with the "relevant social positions" to which Rawls refers without employing some information from our own society. What is problematic is that Rawls focuses selectively on certain factors that determine one's social position, viewing class and economic position as significant and overlooking the ways in which factors like sex also affect one's social power and position. Moreover, by offering an account of "justice" that fails to address the *injustice* of sexism, Rawls seems to overlook the significance of this problem.

Second, it is not clear that the issues of gender and race oppression would in fact be part of Rawls's "nonideal theory." His brief discussions of nonideal theory do not focus on the oppression of women or people of color as central examples. He notes that nonideal theory should be understood as consisting of distinct "subparts": one involves "principles for governing adjustments to natural limitations and historical contingencies," and the other consists of "principles for meeting injustice" (*TJ*, 246). In a subsequent discussion of this second "subpart," that of injustice, Rawls states more specifically what this part of nonideal theory covers:

> It includes, among other things, the theory of punishment and compensatory justice, just war and conscientious objection, civil disobedience and militant resistance. . . . I shall take up but one fragment of partial compliance theory: namely, the problem of civil disobedience and conscientious refusal. And even here I shall assume that the context is one of a state of near justice, that is, one in which the basic structure of society is nearly just, making due allowance for what it is reasonable to expect in the circumstances. An understanding of this admittedly special case may help to clarify the more difficult problems. (*TJ*, 351)

Although Rawls does not explicitly mention gender or race oppression, it is of course *possible* that he meant to include them among the types of injustice with which "nonideal" theory would have to deal. Nonetheless, the sorts of injustice on which he focuses seem rather different from the systematic injustices of gender and race oppression. On the one hand, theories of punishment and compensation typically focus on individuals, and on the other, theories about just war primarily focus on states. Questions about relations between different groups *within* a single society seem to fall between the cracks.

In exploring Rawls's failure to address racial injustice, Charles Mills provides a somewhat different analysis of ideal theory. Mills agrees that Rawls neglects both feminist and antiracist concerns, and he offers a sharp critique of Rawlsian social contract theory. However, Mills suggests that Okin provides a model of how Rawls's theory can be reformulated so that it can address racial and gender oppression (Mills 2007, 2009). The new revised contract, which Mills refers to as the "domination contract," differs from Rawls's social contract in at least two significant ways: (1) it is based not on individualism but on an ontology of social groups, which it derives from an examination of history, in which persons are oppressed based on social group membership, and (2) it is nonideal and as such concerned with matters of corrective justice.

According to Mills, Rawls acknowledges that theorizing can be either ideal or nonideal and simply (and wrongly) focuses his efforts on ideal theory, presupposing but never actually arguing that we must work out an ideal conception of justice before applying it to the nonideal realm of actual problems. Rawls's starting point "handicaps his enterprise," since abstracting away from problems of social subordination works to obscure them, and since problems arise in nonideal theory that are different from those that arise in ideal theory (Mills 2007, 94). Thus Mills concludes that Rawlsian theory must at least be "informed by the non-ideal" (95). However, when Mills writes about the nonideal, what he means by "non-ideal" is really quite different from what Rawls means when he writes about the nonideal as concerning corrective or rectificatory justice. On Mills's understanding, but not on Rawls's, nonideal theorizing requires acknowledging the effects that social power structures have on both projects of theory construction and access to facts about society. While ideal theory abstracts away from oppression and thereby obscures it, "nonideal theory recognizes that people *will* typically be cognitively affected by their social location, so that . . . the descriptive concepts arrived at may be misleading" (Mills 2004, 173). After discussing examples such as the feminist concept of sexual harassment and the Marxist concept of class alienation, Mills notes that concepts "crystallize in part from experience" and that "it may be that the nonideal perspective of the socially subordinated is necessary to generate certain critical evaluative concepts in the first place" (2004, 175). What Mills demonstrates is that ideal theory is insufficient; the answer, however, is not simply that we need to be doing "nonideal theory," but nonideal theory of a certain sort—theory that pays particular attention to the experiences of the oppressed. Thus Mills is not merely plugging in additional information and doing theory that is "nonideal" as opposed to ideal. Rather, his work actually involves a particular attention to the nonideal conditions of oppression and social subordination. Unfortunately, it is precisely this attention to actual social conditions that Rawls's work lacks.

Relevant Social Positions and Knowledge of Oppression

To address his relative silence on the issue of women's oppression, Okin argues that Rawls's theory should be amended to include women as a "relevant social position" in the OP and as a potential group among the "least well off." She explains, "once we challenge Rawls's traditional belief that questions about justice can be resolved by 'heads of families,' the 'least advantaged representative woman,' who is likely to be considerably *worse* off, has to be considered equally" (Okin 1989b, 245). Making a similar point, Eva Kittay suggests that the social positions of both dependents and those who do "dependency work" (most often women, in our current society) should be included as "relevant social positions" in Rawls's OP. Like Okin, Kittay concludes that the group of "least advantaged" citizens should not be defined solely in terms of economic class, but that this group should be understood as including dependency workers (Kittay 1999, 110–12).[2]

One could object to this feminist revision of Rawls by pointing out that it directly contradicts Rawls's own claim that the category "least well off" is not to be defined in terms of gender, which Rawls states clearly in *JFR* even after acknowledging that the parties in the OP should not be understood as heads of families but as persons who are ignorant of their gender.[3] Nonetheless, this is not my objection to Okin and Kittay, since they offer only an interpretation of Rawls and argue that his theory holds potential for feminists, not that it can be applied without revision to bring about gender justice. My objections to the inclusion of gender as a relevant social position in the OP are more fundamental and concern the very meaning of a "relevant social position" in an ideal theory. On the one hand, most feminists would agree that, owing to the impact of sexism, combined with other forms of structural oppression, many women in our society are part of what we would consider the "least well off" social group, as are many of those persons Kittay describes as dependency workers and dependents. Nonetheless, Kittay and Okin seem to be misinterpreting Rawls when they suggest that the "relevant social positions" in the OP are patterned on the actual social positions in our own society. According to Rawls, the OP is a thought experiment whose relevant social positions are generated by thinking reflexively about, and using our considered judgments to determine, what the relevant social positions would be in an abstract ideal, which Rawls calls a "well-ordered society"; they are not determined by looking around our own society and deeming the various social groups that we find "relevant social positions."

Of course, one might claim that when Rawls himself discusses what would count as relevant social positions, he cannot help but consider the social positions of our own society and import these into the OP. To envision what these social positions would be, we must employ our own considered judgments, which are not entirely abstract or free of bias. Nonetheless, Rawls focuses almost exclusively on economic classes, which he claims would continue to exist in a well-ordered society. He argues (however unsatisfactorily) that owing to increased incentives and productivity, some class-based inequalities benefit the least well off and thus would persist in a well-ordered society. Whether or not feminists find these arguments about class convincing, they would certainly object to the continuation of inequalities based on gender. Yet the claim that women, dependency workers, or any other currently oppressed group should be considered a "relevant social position" in the OP seems to imply that inequalities based on gender, dependency status, or some other such feature would continue to exist in a well-ordered society. Such an assumption is problematic for feminists who envision a world in which gender does not determine social status.[4] Interestingly, Okin herself argues that "[a] just future would be one without gender" (*JGF*, 171) and that "the disappearance of gender is a prerequisite for the *complete* development of a nonsexist, fully human theory of justice" (105). Thus Okin posits a category in the OP (gender) that she ultimately believes must be eradicated in a just society. Not only does this go against what it means for something to be a category in the abstract ideal of the OP, but it also has the very troubling implication that gender is a legitimate category of social status and that it will continue to be a marker of inequality in a well-ordered society.

Another way in which Okin departs from the abstract idealization of Rawls's OP is in her discussion of the knowledge possessed by parties behind the veil of ignorance. As I noted above, Okin claims that because Rawls grants that the parties would "know the general facts about human society" (*JGF*, 91), they would know that society "is gender-structured both by custom and still in some respects by law" (91) and "that women have been and continue to be the less advantaged sex in a great number of respects" (102–3). The question of what falls under the "general facts about human society" is a tricky one, as Okin herself explains (191). Rather than offer an argument for her interpretation, however, Okin simply assumes that knowledge of gender oppression would be included in these "general facts." In describing the OP, Rawls provides a list of what would and would not be known to the participants. In short, the parties would be denied knowledge of their place in society, their social status or class position, their natural abilities and assets, their personality and talents, their conception of the good, and their own psychological features. The parties "do not know the particular circumstances of their own society . . . its economic or political situation, or the level of civilization and culture it has been able to achieve" (*TJ*, 137). While they would know the "general facts about human society," Rawls takes this to mean that they would understand "political affairs and the principles of economic theory" and "the basis of social organization and the laws of human psychology" (137).[5]

Looking carefully at this list, it seems that Okin has read too much into the "general facts" that Rawls attributes to the parties in the OP. Facts about current and historical patterns of social domination and oppression—whether racial, gendered, or any other sort—are precisely the type of information that Rawls excludes. According to Rawls, "general facts" encompass only basic, generic information: laws of psychology and principles of social and economic organization that would apply generally and would be widely accepted and relatively uncontroversial. While "general facts" about the oppression of women in current and recent history may be unobjectionable and obvious to feminists, they are controversial and— more important—are not the generic type of knowledge of human society that would be known from the perspective of an abstract ideal. They seem much more akin to the knowledge that Rawls prohibits: facts about the "particular circumstances" of our own society and its "economic or political situation" and culture. Yet without this specific historical information about oppression included in the OP, the thought experiment would not generate the feminist conclusions that Okin desires.

Accommodating "Pluralism" of Views About Gender

Whereas Okin seems to depart from Rawls by including women as a relevant social position and attributing knowledge of gender oppression to the parties in the OP, she sticks closely to Rawls—perhaps too closely—in the matter of accommodating the actual "pluralism" of views found in our own society. Although she believes that a just future would be genderless,

Okin argues, out of respect for the pluralism of current views about gender, that "when we think about constructing relations between the sexes that could be agreed upon in the original position . . . we must also design institutions and practices acceptable to those with more traditional beliefs about the characteristics of men and women, and the appropriate division of labor between them. . . . Gender-structured marriage, then, needs to be regarded as a currently necessary institution (because still chosen by some) but one that is socially problematic" (JGF, 180). I find this element of Okin's work troubling, since it seems to conflate the need to respect a diversity of views with the need to respect the particular diversity of views found in our own society.

In *PL*, where he discusses pluralism more extensively, Rawls argues that any just society will continue to have a variety of different views: this diversity—and specifically the diversity of "reasonable comprehensive doctrines"—is not something that would disappear in a just society (*PL*, 36–37). While Rawls is certainly correct to suggest that a free society would manifest a variety of reasonable views on many different questions of social and political significance, it seems illegitimate to assume that the pluralism of our own society resembles the pluralism that would result from "just" or "free" institutions. Although Rawls does not explicitly state that he means to preserve the actual diversity of perspectives in our own society (in fact, he distinguishes the fact of "reasonable pluralism" from the fact of "pluralism as such"), he often writes as if the current comprehensive doctrines—such as those of the major religions—are acceptable as part of the reasonable pluralism that naturally would develop in free institutions. For instance, Rawls states that "the diversity of reasonable comprehensive religious, philosophical, and moral doctrines *found in modern democratic societies* is not a mere historical condition that may soon pass away; it is a permanent feature of the public culture of democracy" (*PL*, 36, emphasis added).[6]

Okin seems to follow Rawls in sliding from the abstract fact of "reasonable pluralism" to the assumption that the pluralism of views in our own society must be respected. As she explains, "we may, once the veil of ignorance is lifted, find ourselves feminist men or feminist women whose conception of the good life includes the minimization of social differentiation between the sexes. Or we may find ourselves traditionalist men or women, whose conception of the good life, for religious or other reasons, is bound up in an adherence to the conventional division of labor between the sexes" (JGF, 174). However, understanding the OP as an ideal theory, it seems highly problematic simply to import views that are based on the specific social and historical developments of our own nonideal society. Although there would be a diversity of perspectives in the OP, we have no reason to think that this diversity would include "traditional" views about gender. "Traditional" views arose in a particular social context, characterized in part by a history of various forms of oppression; such views are not merely the natural by-products of a free and just society.

Ideal Theory and Reasonable Pluralism

In the move to a purely political form of liberalism, Rawls introduces a number of new ideas: a political conception of justice (as opposed to a comprehensive doctrine), an overlapping consensus, public reason, a political conception of the person, and reasonable as opposed to simple pluralism. While many feminist objections to political liberalism focus on the problematic ways in which issues of importance to feminism, such as the family, child care, reproduction, and sexuality, are relegated to the "nonpublic" and "nonpolitical" sphere and are therefore seen as outside the scope of justice, less attention has been paid to the problematic way in which Rawls's recent work wavers between idealized discussions of a well-ordered society and actual descriptions of our own society. Even if we were to accept the problematic public-nonpublic and political-nonpolitical distinctions, it seems that we still would not know—in the abstract—what sorts of views would develop in the nonpublic and nonpolitical realm of a well-ordered society. Instead of exploring the question of what sorts of comprehensive doctrines persons might develop in a just, democratic society with free institutions, Rawls seems more interested in arguing that the actual comprehensive doctrines that people currently hold are acceptable as part of the "overlapping consensus."

On the one hand, none of these concepts themselves entails the acceptance of the actual plurality of views in our own society. For instance, Rawls explains reasonable pluralism by saying that it is the "normal result of the exercise of human reason within the framework of the free institutions of a constitutional democratic regime" (*PL*, xviii). On the other hand, although this explanation sounds abstract and makes no explicit reference to our own society, as Rawls fleshes out his motivations for writing *PL*, it becomes clear that he is in fact talking about our own society. Consider the following passage: "A modern democratic society is characterized not simply by a pluralism of comprehensive religious, philosophical, and moral doctrines but by a pluralism of incompatible yet reasonable comprehensive doctrines. No one of these doctrines is affirmed by citizens generally. Nor should one expect that in the foreseeable future one of them, or some other reasonable doctrine, will ever be affirmed by all, or nearly all, citizens" (xviii). Here Rawls implies that actual "modern democracies" are in fact the background context for his own discussion of pluralism. In other words, he suggests that our own modern democratic society is characterized by reasonable pluralism, and the problem for political liberalism is to explain how it is that we can maintain most forms of pluralism (with the exception of some extreme views that are "unreasonable," such as religious fundamentalism) and still agree on the political ideal of "justice as fairness." Although Rawls claims to be doing abstract theory, it seems that his intention in *PL* is to show how the "pluralism" of our own society—especially religious and moral pluralism—is compatible with justice, despite initial appearances to the contrary.

Thus one problem with the way that Rawls presents and discusses the issue of pluralism is that his blurring of the ideal and the actual makes it difficult for one to object to the specific forms of pluralism of our own society without objecting to the very idea that a plurality of reasonable comprehensive doctrines would result from the normal functioning of human reason under free and democratic institutions. Clearly, most feminists do not object to this abstract point; we have no reason to believe that everyone would think alike or would hold the same values and commitments in a just and free society, nor do we want to promote a uniformity of viewpoints. Yet believing that a just society would be pluralistic does not entail support for the current pluralism of values and perspectives in our own society, which has a history and a present marred by various forms of social, political, and sexual oppression.

Initial Social Positions as "Arbitrary"

I have argued that Rawls focuses his discussion of inequality too narrowly on economic issues, that his methodology is not amenable to the addition of women as a relevant social position, that his theory denies parties in the OP information about issues of oppression, and that his defense of reasonable pluralism ultimately works to justify the continuation of current social structures rather than to promote a just social ideal. In addition to these problems, and in a sense underlying them, is a more basic problem with how Rawls conceptualizes social power and inequality. Rawls's discussion of inequality seems to focus on mitigating the effects of the "arbitrariness" of natural advantage and disadvantage rather than on altering the structures of power that generate and perpetuate various forms of social hierarchy and oppression.

Several times in his writings, Rawls describes the "initial social position" into which one is born as "arbitrary," in much the same way that the natural talents and abilities that people are born with are arbitrary. In fact, Rawls frequently lumps together the "distribution of natural talents" and the "contingencies of social circumstance" without acknowledging that there may be important differences between them (*TJ*, 102–3). In this way, Rawls seeks to separate the effects of one's choices from the effects of one's endowments and circumstances, since people cannot be held responsible for their natural endowments or for the social circumstances into which they are born. From Rawls's perspective, both natural endowment and social circumstance seem arbitrary; it is a matter of luck whether one is born with certain natural endowments, just as it is a matter of chance whether one is born into a wealthy or a poor family. He explains that "the natural distribution [of talents] is neither just nor unjust; nor is it unjust that men are born into society at some particular position. These are simply natural facts" (*TJ*, 102).[7] Viewed from the perspective of the individual agent, these may seem like "natural facts" and may appear entirely arbitrary.

Yet when considered instead from a larger perspective, the issue of social circumstance is anything but natural or arbitrary. The social and economic forces that create and maintain structures of power and oppression—including class, race, and gender—are not governed by luck or accident.[8] For instance, it is not merely a matter of luck or chance that men have more power than women, that whites are the dominant race in the United States, and that heterosexuals have more rights and privileges than gays and lesbians do. These are social facts rooted in history and supported by intricate webs of power and privilege. Lumping the effects of socially generated hierarchies with the effects of natural endowment makes it seem as if these structures were not produced by social and political forces. Thus, while it is laudable that Rawls does not merely want to accept whatever effects come about from one's being born into a particular social position, his theory does not go far enough in identifying and challenging the social structures of power that generate these initial positions.

Conclusion: Social Injustice and the Limitations of Rawlsian "Abstraction"

In contrast to Rawls's focus on the perspective of individuals, feminist discussions of justice and equality typically center on relationships between social groups and on the power structures that generate unjust hierarchies. Remedying social injustice requires attention not only to the differences between individuals at any given point in time but also to the social forces and relations of power that perpetuate and sustain these differences. As Iris Marion Young argues, justice is not merely a matter of "distribution," and it cannot be fully understood apart from the institutional context in which patterns of distribution are created and sustained.[9]

On the one hand, it might seem as if Rawls's discussion of the importance of opportunities and self-respect, as well as his claim that people deserve the "fair value" of their equal political liberties, would work to ensure basic equality in the larger culture. Despite this appearance, however, both of his major books, as well as his essays and subsequent writings, fail to describe in much detail the cultural and social forces that create and sustain social hierarchies. For Rawls, power dynamics in the "background" culture are beyond the scope of his theory, since his theory allegedly concerns only the "basic structure" of society.[10] The ideas and values that people hold are typically understood to be matters of their own personal perspectives—such as their moral views or their religious beliefs—and are not considered "political." Thus many elements of culture and institutional context are relegated to the periphery and are seen as separate from the "political" conception of justice Rawls offers. Instead of challenging the cultural and institutional context that generates and perpetuates various forms of oppression, Rawls leaves this context aside and focuses on remedying the unfair advantages and disadvantages associated with various social positions within this context.

Thus, although Rawls claims to be offering a theory of justice based on an abstract ideal of a just basic structure, there are a number of ways in which his "ideal" fails to question, and even implicitly assumes, the power structures of our own society. On the one hand, there are several ways in which gender is absent from Rawls's discussion: he does not include women as a relevant social position, nor does he grant the parties in the OP knowledge of gender oppression. While some feminists (such as Okin) suggest that these problems with Rawls's theory of justice can be remedied by including women in his theory, I conclude that the problems with Rawls are more fundamental and are linked to an underlying failure to question—and even to identify—socially created and maintained structures of hierarchy. Feminists argue that male dominance is not a natural outgrowth of men's and women's "differences." To challenge sexist structures, we need to do more than compensate women for the "disadvantage" of being born women in a sexist society. We must look more deeply at the causes of women's oppression and at the structures of power that create and sustain it.

While my criticisms aim to undermine Rawls's claim to abstraction, I do not mean to suggest that political theory simply needs to be more "abstract." Feminist change requires envisioning some other alternative society—one where the institutions of the basic structure do not take gender oppression, or other forms of unjust hierarchy, as a given. Such radical envisioning requires a careful contextual analysis of the mechanisms of domination and oppression in our own society, and this sort of analysis seems to fall well outside the method and framework Rawls proposes. Thus, while critical attention to Rawls's work may be useful in understanding some of the problems with contemporary philosophical approaches to justice, feminists ultimately must move beyond the confines of Rawlsian abstraction to construct theories that illuminate current relations of power and privilege and that envision more radical alternatives.

Notes

1. I thank Ann Cudd for urging me to consider this line of argument.

2. Other feminist analyses of Rawls draw similar conclusions, arguing that the social positions of men and women both need to be considered by the parties in the OP (Green 1986; Trout 1994), but I focus on Okin and, to a lesser extent, Kittay.

3. Rawls explains that "in the simplest form of the difference principle the individuals who belong to the least advantaged group are not identifiable apart from, or independently of, their income and wealth. The least advantaged are never identifiable as men or women, say, or as whites or blacks, or Indians or British" (JFR, 59n26).

4. Similarly, in terms of dependency work, an ideal "well-ordered" society might involve the elimination of *classes* of people who do this work; it is at least possible to imagine that the work of caring for dependents would be much more evenly spread throughout the population, or that certain people would engage in dependency work for short periods of time and yet not have their social role defined by it in the way that many people (and, specifically, many women) currently do.

5. Deborah Kearns claims that it is very unlikely that persons in the OP would agree on subjects as controversial as those that Rawls lists and notes that "besides diversity at any one time, historically the leading principles of economic theory and the dominant understanding of social organisation have changed" (1983, 37).

6. Rawls claims explicitly that he "counts many familiar and traditional doctrines—religious, philosophical, and moral—as reasonable even though we could not seriously entertain them for ourselves" (PL, 59–60).

7. Rawls does go on to argue that there are just and unjust ways of dealing with these "contingencies," even though he does not explain the structure of positions itself as being a matter of injustice.

Although he aims to change the social system, he still suggests that the positions into which we are born are "accidents of social circumstance" rather than the products of socially created structures.

8. For an insightful critique of "luck egalitarianism," see Anderson 1999.

9. See Young's (1990) discussion and criticism of the distributive paradigm. She contends that this paradigm treats nonmaterial social goods as if they were material things to be possessed. While it is important to note differences in the rights, opportunities, and self-respect of individual persons, the suggestion that these are things to be possessed distorts the relational character of these "goods."

10. As discussed in the introduction to this volume, Rawls attempts to clear up questions about whether the family is part of the "basic structure" (or whether it is merely an "association") in IPRR. For an insightful analysis of Rawls's failure to answer feminist criticisms of his treatment of the family, see Smith 2004. See also Abbey 2007 and chapter 4 in the current volume.

3

Rereading Rawls on Self-Respect

Feminism, Family Law, and the Social Bases of Self-Respect

Elizabeth Brake

In *TJ*, Rawls famously wrote that self-respect is "perhaps the most important primary good" (440). Indeed, he adduced its primacy to defend his theory of justice in a number of places. But as subsequent discussion has shown, he was imprecise in defining "self-respect," using the term interchangeably with "self-esteem." Moreover, he failed to examine systematically the implications of its primacy; while he claimed that equal liberties, equal opportunity, the difference principle, and the contract model itself supported self-respect, he did not inquire whether there were other impor-

---

For helpful comments, I wish to thank the audiences at the 2011 annual conference of the Society for Applied Philosophy, the University of British Columbia Philosophy Department, the Arizona State University Philosophy Department, the 2010 annual meeting of the American Political Science Association, and, especially, Ruth Abbey.

tant social determinants that should, in consistency, be treated as social bases and distributed equally. In other words, he adduced the primacy of self-respect in key places to buttress his theory, but he did not consider the implications of its primacy as an independent topic. This inattention to the implications of the primacy of self-respect has left his theory open to the charge that it requires too much, but it also suggests an untapped resource for feminist interpreters of Rawls.

In this chapter I first consider the argument for the status of self-respect as a primary good and the problems raised by Rawls's slippery definition; I defend a definition that, although it goes against the grain of much Rawls interpretation, serves the theory better. I then turn to the implications for family law of the status of self-respect as a primary good. I focus on the family because it is a crucial site of the development of self-respect and because its effects on female children's life chances are of particular interest to feminism. Okin calls the gender-structured family the "linchpin" of women's oppression, disadvantaging women through the gendered division of labor and the inculcation of gender roles (JGF, 6). I argue that the status of self-respect as a primary good has implications for the legal recognition of family forms and for parents' rights to "infuse" children with their beliefs. While critics have objected that the implications of the status of self-respect as a primary good are too demanding, I argue that its demands suggest the potential for Rawls's theory to address women's oppression at the root.

A few examples will help suggest the main line of thought. Okin argued that indoctrinating girls into sexist religious practices could jeopardize their self-respect, asking, "How can a small Catholic girl, for example, who identifies with the priest and seeks to emulate him in her future life, be expected to retain her self-respect . . . on being told that it is impossible for her to become a priest, simply because she is a girl?" (2005, 242). Self-respect can be imperiled by the transmission of secular as well as religious beliefs, as when a daughter is taught that her value rests solely on her looks. Self-respect can also be vitiated through mere treatment, as well as explicit indoctrination. In Jane Smiley's novel *A Thousand Acres*, the protagonist, Ginny, is subjected to childhood abuse and neglect and consequently becomes a woman so afraid of displeasing others that she is unable even to entertain her own opinions. She seems not to understand that women are entitled not to be abused by their husbands. She describes her father as treating female children as an owned resource, like land and animals on his farm. His treatment of her has caused her to internalize this view of herself. Not only does she lack a sense of entitlement and legitimate boundaries; she lacks the internal resources to maintain her own viewpoint. Ginny, in short, lacks self-respect, and this prevents her from forming and acting on a conception of the good.

The following argument for legal frameworks that protect children's self-respect does not systematically investigate the social bases of self-respect. My point is that Rawls ignored some important social bases of self-respect, and that the status of self-respect as a primary good requires greater intervention in the family than Rawls realized. Nor does my focus on the family imply that there are no other influences on women's self-respect, such as images of women in pornography and the media, women's political representation, and sexual harassment. The status of self-respect as a primary good has implications in many areas of law.

Self-Respect as a Primary Good

As noted above, Rawls deems self-respect "perhaps the most important primary good." He defines primary goods in *TJ* as goods needed regardless of one's particular conception of the good, goods that any rational person will want, and all-purpose means for any plan of life. Primary goods fall into two groups: the social, which society can distribute directly, and the natural, which society influences but cannot distribute directly. Wealth and income, opportunities, and liberties are examples of the former; health, intelligence, imagination, and vigor, of the latter (*TJ*, 62). Their definitive feature is that they serve the pursuit of any plan of life.

Primary goods are a means of comparing citizens' positions; they are the goods that the principles of justice distribute. As such, they must be measurable and distributable. Primary goods are an objective standard and are contrasted with subjective standards such as welfare; their objectivity facilitates interpersonal comparisons and holds citizens responsible for their tastes—for example, by a subjective standard, the indulgent gourmand might require more resources than the Spartan eater.[1] Further, primary goods are supposed to bridge the competing conceptions of the good found in society. Rawls's "thin theory of the good" is intended to take a neutral position on competing conceptions in order to avoid illegitimately favoring a particular view of the good.[2]

Rawls writes that self-respect is "perhaps the most important primary good" because it is a psychological precondition for the pursuit of plans of life:

> We may define self-respect (or self-esteem) as having two aspects. First of all . . . it includes a person's sense of his own value, his secure conviction that his conception of his good, his plan of life, is worth carrying out. And second, self-respect implies a confidence in one's ability, so far as it is within one's power, to fulfill one's intentions. When we feel that our plans are of little value, we cannot pursue them with pleasure or take delight in their execution. Nor plagued by failure or self-doubt can we continue in our endeavors. It is clear then why self-respect is a primary good. Without it nothing may seem worth doing, or if some things have value for us, we lack the will to strive for them. All desire and activity becomes empty and vain, and we sink into apathy and cynicism. Therefore the parties in the original position would wish to avoid at almost any cost the social conditions that undermine self-respect. (*TJ*, 440)

The general idea has intuitive appeal: in Smiley's novel, Ginny's lack of self-respect causes her inability to choose and pursue her own goals because she lacks any sense of entitlement. Rawls adduces the primacy of self-respect in favor of the principles of justice, their lexical ordering, and the contractarian model itself (Eyal 2005, 196). He grounds a crucial argument for the lexical ordering of his principles of justice on the claim that equal liberties are the social bases of self-respect (*TJ*, 544–48). He writes in favor of the liberties that they symbolically support self-respect by representing citizens as equals, as well as by securing the right to participate in associations that support self-respect (233–34). He defends the contractarian approach as supporting self-respect by representing citizens as equals, and he defends the difference principle against the objection that inequalities will undermine self-respect (545–46). The primacy claim is important to Rawls's overall argument. But is his argument that self-respect is a primary good at all, perhaps even the most important primary good, successful?

The answer turns, in part, on the much debated definition of self-respect. To complicate matters, Rawls revises his account of primary goods and self-respect in *PL*. There he grounds the account in the political conception of the person: primary goods are means normally needed to develop and exercise the moral powers, namely, the capacity for a sense of justice and for a conception of the good (*PL*, 75–76). These powers are, in Rawls's theory, the basis of moral equality, and they define the political conception of the person.[3] Since justice as fairness responds to the moral significance of citizens' capacity to choose their life plans, there is a close fit between the fundamental normative concept in the theory and the definition of primary goods in terms of the pursuit of life plans. Further, in *PL* self-respect involves seeing oneself as a free and equal citizen among others, possessing the moral powers, and as a "fully cooperating" member of society (76–77, 81). However, most debate over Rawls's definition of self-respect has focused on the fact that he does not distinguish self-respect and self-esteem in *TJ*.

Critics have argued that there is a distinction between self-respect and self-esteem, and that Rawls's argument in fact implicates self-esteem, understood as a positive self-evaluation or as "confidence in one's determinate plans and capacities" (Eyal 2005, 202). One reason for reading his argument as concerning self-esteem is the view that self-respect cannot be shown to be a primary good (Eyal 2005, 203–5). I will argue that self-respect is the preferable notion, and that self-respect can be shown to be a primary good. Self-esteem and self-respect differ in at least two important ways. Self-esteem has broader bases than self-respect; it may be bolstered, and diminished, by things as disparate and irrational as appearance or relaxation exercises. Self-respect is a response to one's worth as a certain kind of being, and it is essentially normative, although it may be undermined irrationally. In addition, self-esteem is evaluative, whereas self-respect need not be.

Stephen Darwall's distinction between "recognition respect" and "appraisal respect" is useful in clarifying the second point (Darwall 1977; cf. Middleton 2006; Doppelt 2009). Appraisal respect, as Darwall defines it, involves evaluation of a person's positive qualities, such as virtues or abilities, and varies in amount as the qualities vary. Darwall contrasts appraisal respect with recognition respect, which responds to the object of respect as a kind of being deserving certain consideration, such as a rational agent or a moral person, and attaches equally to all such beings. Thus when Nir Eyal writes that Rawls intended a Kantian notion of self-respect—"confidence that one has the dignity of persons"—he attributes to Rawls a recognition notion of self-respect (Eyal 2005, 203). Adapting Darwall's terms, I will henceforth distinguish between *appraisal self-respect* (which I extend to include self-esteem) and *recognition self-respect*.

Most commentators take Rawls's argument in *TJ* as requiring appraisal self-respect because it seems to hold that self-respect involves evaluation of one's abilities (Eyal 2005, 206; Moriarty 2009, 455). In the passage cited above, Rawls writes that self-respect involves "confidence in one's ability, so far as it is within one's power, to fulfill one's intentions." Because this seems to require evaluation of one's abilities, commentators have taken Rawls as (perhaps inadvertently) endorsing appraisal, not recognition, self-respect. But this creates problems for the theory. As Gerald Doppelt writes, appraisal self-respect is too "precarious, variable, and subjective for Rawls' purposes" (2009, 133–34). If self-respect is a primary good, presumably its social bases should be distributed equally. But because appraisal self-respect allows comparisons, it would be vulnerable to the inequalities permitted by the difference principle. This prompts Eyal to mount a *reductio* objection to its primacy on the grounds that distributing the social bases of appraisal self-respect equally would require strict egalitarianism and curtailment of liberties.

Moreover, not only is appraisal self-respect affected by many and disparate things, so that distribution of its social bases is in tension with our considered moral judgments, but its determinants would include whatever normally improves agents' self-evaluation—such as Prozac or psychotherapy—even if this distorts their judgment! Confidence in one's abilities may be unjustified. Further, appraisal self-respect might suffer from similar problems as do subjective accounts of the good. Just as a utility monster might create unfair distributions by requiring disproportionate resources, an appraisal self-respect monster might create unfair distributions by requiring *favorable* comparisons with others.

Recognition self-respect is the preferable notion because it avoids the problems of comparative and subjective evaluations. It also fits more closely with the grounding of the primary goods in the political conception of the person. On the political conception, the person is characterized by a capacity to have a conception of the good—not a high self-evaluation. Recognition self-respect does not require an evaluation of the objective worth of one's plan or of one's current abilities; in the context of Rawls's theory, it requires a recognition of oneself as an agent possessing the moral powers and, consequently, as having certain entitlements (PL, 76–77). But can the primacy argument succeed on this interpretation of self-respect? That is, does recognition self-respect enable us to carry out our plans? To show that it is, at least, a primary good, one must show that it is normally needed to carry out one's plans, whatever those may be.

In the passage cited above, Rawls distinguishes two ways in which self-respect contributes to the pursuit of plans. The first concerns "a person's sense of his own value" and the worth of his plan. The sense of one's own value is clearly a normative notion, comporting with recognition of oneself as a free and equal citizen possessing moral powers. But a person's sense of the worth of her plan appears to be an evaluative notion. However, an alternative reading is possible. Rawls foregrounds the agent's belief in her own value: "a person's sense of his own value, his secure conviction that his conception of his good, his plan of life, is worth carrying out" (TJ, 440).

There is an ambiguity parallel to that in Plato's *Euthyphro*: does the agent view her plan as worth carrying out because it is objectively and independently valuable, or because it is *hers*? If the former, then the belief that her plan is worthy is primarily an evaluative judgment about the plan. But such a judgment is not a judgment about oneself at all, and hence not part of self-respect! Rawls could not consistently hold that we derive our value from our plans, because his account of our interest in liberty holds that the value of our ability to choose is prior to and independent of the plans we choose (*PL*, 30). Thus the latter reading is preferable. On this reading, the judgment is about the agent herself—her choosing the plan makes it worth pursuing, and in this respect she is equal with other such choosers. Recognition self-respect supports the pursuit of plans by warranting action: it implies that my plans give me reason to act; they are reason-giving simply as my plans, and they are as reason-giving for me as others' plans are for them.[4]

The second way in which self-respect supports the pursuit of life plans appears to concern empirical self-evaluation of one's abilities: without "confidence in one's ability, so far as it is within one's power, to fulfill one's intentions," one would not undertake anything, beset by "self-doubt" and "lack[ing] the will to strive" (*TJ*, 440). In a rational person, there is a close connection between confidence in one's abilities and undertaking plans: within the rational choice theory that Rawls assumes in *TJ*, a rational individual will order her preferences according to her options (*TJ*, 407–16). If I cannot become a professional basketball player, it would be irrational for me to spend my life pursuing this goal. This aspect of self-respect supports the pursuit of plans because it involves beliefs about abilities that make such pursuits rational.

But these beliefs *cannot* be empirical self-evaluations, and hence the relevant notion cannot be appraisal self-respect. If the argument required that the agent perform empirical self-evaluation of current abilities, self-respect would require that epistemically finite agents acquire false beliefs, restricting self-respect to the extraordinarily capable or delusional.[5] Suppose someone wants to make a philosophical contribution or combat global warming but doesn't know whether she will be able. Perhaps she's considering graduate school or volunteering for Greenpeace. Uncertainty concerning one's ability to carry out ambitious or long-term projects is common. Ambitious and long-term plans challenge the appraisal self-respect reading because the relevant knowledge is unobtainable. There are many projects that may be close possibilities for a given agent, without her being able to evaluate whether her abilities will meet the challenge.

Abilities are malleable and responsive to training; one can develop them beyond what is expected. For many purposes, their extent is not determined. A person's abilities depend on education and social supports, but also, significantly, on her own efforts. Belief in one's agency and potential may indeed be the most crucial determinant of development, since without it one may not even make an effort. This, I take it, is Rawls's point.

Rather than empirical self-appraisal, "confidence in one's ability . . . to fulfill one's intentions" can be understood as a recognition of one's agency, one's status as the kind of being who can, by willing, cause results, and thus whose striving is worthwhile. This belief falls under recognition, not appraisal, self-respect because it concerns the recognition of one's status as an agent, not the empirical appraisal of one's abilities. Given our epistemic limits as finite rational agents, recognition self-respect serves the argument better than appraisal self-respect does.

Recognition self-respect avoids the problems with appraisal self-respect.[6] It does not require that the agent develop false beliefs. It is more robust in the face of socioeconomic inequalities—whereas, because appraisal respect involves comparisons, it might be more vulnerable to the inequalities permitted by the difference principle. Recognition self-respect fits more closely with the account of the primary goods and the political conception of the person, and it explains Rawls's comments about the agent's confidence in her plan, which otherwise appear not to be properly part of self-respect. In sum, Rawlsian self-respect should be understood as recognition of oneself as an agent possessing the moral powers and, accordingly, recognition that these powers give one entitlements as an equal citizen.

On this definition, is recognition self-respect a primary good? The beliefs that choosing a plan gives one a reason to pursue it, and that one's willing will be efficacious, seem fundamental to action; without them, directed action seems unintelligible. Even for someone committed to masochism or servility, his pursuit of such plans seems to require at least the belief that his preference matters and that his agency can be efficacious. This interpretation of self-respect does not seem to capture all that Rawls had in mind; but it serves the argument that self-respect is a primary good—an all-purpose means for pursuing plans of life—while avoiding the vulnerability of appraisal self-respect to *reductio* objections.[7] Rawls apparently thinks that it is "perhaps the most important" because without it, other primary goods would be useless; it is a precondition of pursuing plans at all. However, I have not argued for its primacy but simply for its status as a primary good.[8]

This account can answer a second objection to recognition self-respect. Self-respect might support the pursuit of life plans rationally or irrationally, that is, either by giving reasons to act or merely by motivating action. Eyal argues that appraisal self-respect *warrants* action by giving reasons to act, while recognition self-respect (seeing oneself as possessing dignity) is not appropriately reason-giving (Eyal 2005, 203–4). The self-respect that is a primary good should be *reason-giving*, not merely a psychological spring to action. One reason for this in Rawls's theory is his exclusion of *unreasonable* envy, and its effects on self-respect, as a reason against the difference principle (*TJ*, 546–48). Moreover, if self-respect simply provides a psychological trigger for action rather than a reason for it, its social bases might include motivational tapes and posters, coffee or other stimulants, invigorating music, and other subrational motivators.

However, the foregoing account of recognition self-respect does explain how it *warrants* action. First, it includes a basic assumption of instrumental rationality: one's plans give one reason to act, simply as one's plans. Second, it emphasizes the equality, in this respect, between each person and others; without some special reason (such as a contractual obligation), my plan should not be subordinate to that of another. Finally, belief in one's agency warrants action, under conditions of uncertainty regarding the extent of one's abilities that actual choosers face; that one can cause effects by acting is another fundamental assumption underlying rational action.

If self-respect is an important primary good, justice presumably requires that its social bases be distributed equally. These social bases will be socially distributable determinants of self-respecting beliefs, that is, beliefs that one possesses the moral powers and is a free and equal citizen. These social bases are "determinants" reasonably connected to its possession, not necessary conditions (Moriarty 2009, 443). Someone reared in a fascist dictatorship may develop self-respect without liberties. But the equal liberties will tend to support self-respect, and their symbolism gives citizens reason to think of themselves as equals. The social bases of self-respect may support self-respect either materially or symbolically. For example, equal liberties support self-respect *materially* by securing rights to engage in activities that support self-respect, and *symbolically* by representing citizens as equals. Symbolism is not enough: equal liberties will not support self-respect if the individual is universally reviled and treated as a second-class citizen, or if he is unemployed in a society that connects employment to worth, or if his agency is blocked at every turn. If severe poverty in an affluent society destroys self-respect, a social minimum will be an economic basis of self-respect. Indeed, Rawls suggests correcting large distributive inequalities to avoid harms to self-respect (*TJ*, 545–46; see also Doppelt 2009).

However, distributing the social bases of self-respect equally does not imply strict egalitarianism in the distribution of wealth. Above a threshold, more money does not support recognition self-respect; because it does not vary in degree, recognition self-respect is compatible with some social and economic inequalities. Indeed, there is further reason to think that the difference principle is compatible with recognition self-respect: striving and succeeding (as opposed to having things handed to you) confirms one's agency for oneself. This suggests that job training and job creation as a social basis for self-respect, like finding and succeeding in work, can be a crucial confirmation of agency (Shiffrin 2003–4, 1668; Moriarty 2009).

In *TJ*, Rawls stresses "the great significance of how we think others value us" (544; cf. 441–42); in particular, he emphasizes the supportive effects of interaction with like-minded others. Empirical research supports this point: for example, research on the children of lesbian parents shows a protective effect on such children's self-esteem when they associate with other children of lesbian parents (Bos and Van Balen 2008). It is reasonable to think that this protective effect extends to recognition self-respect: we develop our self-understanding in interaction with others and through internalizing cultural norms and judgments. Our sense of our worth and entitlements depends to a great extent on what is reflected by others and our society. However, sometimes supportive associations are not available, as for a gay teenager in a rural area, or are unable to counteract detrimental abuse in the family, or themselves inflict harm. The words and actions of others may contradict as well as confirm the beliefs involved in recognition self-respect.

Implications for Family Law

Recognition self-respect involves beliefs that can be undermined by the words or actions of others. Rawls unrealistically assumes that agents can avoid speech that undermines self-respect and that the symbolism of the liberties will counteract it. If we grant for the sake of argument that adults may justly choose to expose themselves to such speech, we may construe it as gambling with self-respect, imprudent but not unjust. However, there is a large group of people whose self-respect is exceptionally vulnerable to abuse, people who in many cases lack the power to avoid it: children. I turn now to family law in order to examine the implications of the status of self-respect as a primary good in that domain.

Equal distribution of the social bases of self-respect has implications for legal family recognition and for parental rights. First, where a state recognizes families in law, such recognition is a social basis of self-respect. Second, protecting children's developing self-respect limits parental rights to infuse children with ethical and religious beliefs.

As indicated in the introduction to this volume, Rawls acknowledges that even in a just society, differences in family practices will influence children's life chances unequally (*TJ*, 301). Some parental practices will equip children for future competition better than others; some practices will greatly inhibit children in later pursuits. Baby Einstein DVDs and playing Mozart to the fetus are only extreme examples; most parental practices, such as feeding, play, schooling, and kind or harsh treatment, will have lasting effects on children's life chances. Thus it seems that private parenting may impede societal justice. Rawls was not too concerned about this, because he thought that economic redistribution would offset such inequalities (*TJ*, 544–45). But this is unsatisfying when we consider that even if redistribution prevents significant inequalities in income, differences in parental coaching may still significantly affect children's chances of attaining meaningful work and positions with social status. Furthermore, feminists such as Okin have argued that the gender-structured family, with its gendered division of household labor and teaching of gender roles, is a source of significant inequalities for women. Women who take on the bulk of housework and child care have less competitive advantage in the workplace than their male counterparts. Anticipation of motherhood, when it is seen as incompatible with a full-time career, may limit young women's aspirations. And socialization into hierarchical gender roles begins in childhood.

The family is a major social institution, and as such its outcomes should not be unjust—which is exactly what unjustified inequalities in primary goods such as self-respect are. The status of self-respect as a primary good has revolutionary implications for family law. It implies that damage to self-respect is itself a reason to constrain parental practice, even if such damage does not lead to further inequalities in wealth, income, and position. It thus suggests a Rawlsian strategy for criticizing gender roles that undermine female self-respect directly, and not only because of their socioeconomic outcomes. It explains why legal frameworks should protect women's self-respect, even if damage to self-respect were compatible with distributive justice.

Self-respect, I have argued, should be understood as the recognition of oneself as an equal possessor of the moral powers. We can begin to delineate its social bases by focusing on what would reasonably undermine these beliefs. It is easier to see what is reasonably likely to destroy self-respect than what will encourage it (although Rawls emphasizes participation in associations with like-minded others, as noted above). Recall the first aspect of self-respect: the sense of one's equal value and the belief that one's choices give one a reason to pursue one's plans, as others' choices equally give them reasons. This is reasonably undermined by the belief that one is not valuable, or that one's preferences are not equal reasons for action compared with those of others. Children can develop beliefs that they are not equal sources of value, that their preferences are not equally reasons for action, if their preferences are routinely ignored or subordinated arbitrarily to those of others. A girl who is given second-rate toys and demanding chores while her brother is indulged in every wish would reasonably internalize such an attitude. Children can also develop such beliefs if they internalize the understanding of themselves as inferior types of persons. Shame about one's body, as being the body of an inferior kind of person, can deliver a blow to self-respect. If a girl sees her secondary sexual characteristics as the attributes of an inferior person—a helpmeet, for instance, or a sinful or unclean being—then her sense of herself as equal may be diminished. The same is true of shame about the body's desires and urges, as for sex or food, where these are taken to characterize the person as, for instance, a "slut" or a "fat girl." Negative body image has profound negative psychological effects on adolescent girls (Clay, Vignoles, and Dittmar 2005), and these effects could plausibly extend to the sense of self-worth. The second aspect of self-respect, belief in one's agency, is reasonably undermined by the belief that one's agency is blocked or limited because of the kind of person one is, for example, a girl's belief that she lacks the independence or authority to make her abilities work toward her goals. Protective measures against such undermining beliefs are an important social basis of self-respect. As noted above, symbolism is important: legal structures should not reinforce beliefs about the inequality of persons. But, as I argue below, symbolism is not enough to protect children's developing self-respect.

Let us first consider a case of symbolic equality. Equal distribution of the social bases of self-respect requires the recognition of same-sex marriage and nontraditional families. People tend to be defined by their desire for same-sex sexual partners (as if such desire exhausted their identity) and categorized as "homosexuals" or "bisexuals."[9] Given cultural assumptions about the role and value of marriage, the right to participate in it *as the kind of person one is* (or understands oneself as) is a social basis of self-respect. Especially where the value of marriage is promulgated through educational policies and public pronouncements, and is explicitly linked to good citizenship, as in the United States, the same-sex marriage bar does not simply deprive gays and lesbians and bisexuals of the legal entitlements of marriage; it marks them as excluded from an institution thought to be fundamental to public morality and good character.[10] If the exclusion cannot be grounded in public reason, then the symbolic inequality may reasonably affect an individual's self-respect.

Rawlsian defenses of same-sex marriage have been made on other grounds, primarily on grounds of equality and neutrality. Rawls himself writes that claims against same-sex marriage per se cannot be given in public reason—although, he writes, the effects of same-sex marriage on children could provide public reason against it (IPRR, 157n60). But the current argument speaks to a child-welfare interest in recognizing same-sex marriage: the effects of exclusion on the self-respect of children of gays and lesbians. Restricting marriage to different-sex couples not only affects the self-respect of adult gays and lesbians; it affects the developing self-respect of any children they might have. Discrimination directed at parents affects their children, and the self-respect of children reared within families that lack legal recognition is reasonably affected by the stigmatization and marginalization of their families. A recent study found a "growing consensus among researchers that in terms of psychological adjustment there are no differences between children in planned lesbian families . . . and those raised in heterosexual families," but it reported that experiences of stigmatization by children of lesbian parents (such as bullying, gossip) on the basis of their family structure is correlated with lower self-esteem (Bos and Van Balen 2008).[11] State recognition of such families would counteract stigma, providing symbolic equality. Indeed, this point applies to many "alternative" family forms, such as adult friends communally rearing children (see also Brake 2010).

But social bases of self-respect require more than formal equal rights (Doppelt 2009, 135–42; Moriarty 2009; Shiffrin 2003–4). For instance, while legal equality for lesbian mothers may ameliorate schoolyard bullying, the stigma may persist. The United States has seen such an arc with racism, where the law secures formal legal equality but racist name-calling by children persists. The expression of negative attitudes by others influences children's developing self-respect. For example, peer expression of negative attitudes toward gay and lesbian adolescents in a school's culture, or the unrealistic representation of women's bodies' inducing negative body image in young women, strongly correlates with lowered self-esteem in the affected populations (Wilkinson and Pearson 2009; Clay, Vignoles, and Dittmar 2005). It is plausible to assume that there are similar effects on recognition self-respect.

Unequal treatment or the internalization of beliefs regarding inequality can undermine a child's nascent sense of her equal worth. While political liberals generally tolerate unequal roles in private, given a context of public equality, children cannot make this distinction. A child who is consistently treated as valueless may internalize a belief that she lacks value; a child who is consistently treated as less important than her brother may similarly internalize a belief that she is inferior, which conflicts with the sense of equal worth essential to self-respect.

Mill writes in *The Subjection of Women* that the family is a "school of despotism" when it teaches boys that they possess an arbitrary sex privilege (Mill 1998, 47): "Think what it is to be a boy, to grow up to manhood in the belief that without any merit or any exertion of his own, . . . by the mere fact of being born a male he is by right the superior of all and every one of an entire half of the human race" (86–87).[12] The corresponding impact on little girls must be a belief in their own inferiority, and hence the deprivation of self-respect. Political liberalism might seem to require that little girls develop a dual sense of themselves as equal citizens who are nevertheless unequal in another crucial respect, and this is the distinction that, I am claiming, young children cannot make.

Rawlsian liberalism implies that such outcomes are unjust owing to the status of self-respect within the theory. The deprivation of children's equal self-respect is an injustice to those children, and it leads to further injustices. As children and adolescents, citizens develop aspirations and make crucial decisions—to work hard in school, to direct their efforts to success—that will significantly affect their later life chances. Inequalities in self-respect at this stage can lead to unjustified socioeconomic inequalities later, as when a child's lack of self-respect impedes her studies. Protecting self-respect in childhood seems crucial to children's later life chances in the way that education and health care are. Damage to children's self-respect does not end with childhood. It is highly plausible that in some cases the internalization of such feelings affects the self-respect of the adults whom these children become.

The question is not whether political liberalism should regard damage to self-respect as an injustice, but what it requires in response to such injustice. Interfering with private expressions of views that threaten self-respect, especially within the family, poses challenges. Once again, political liberalism appears to tolerate inequalities and hierarchies, as in religion or intimate relationships, so long as they do not arise from or create injustices. Rawls himself, despite taking women's equality as a public interest, writes that "a liberal conception of justice may have to allow for some traditional gendered division of labor within families—assume, say, that this division is based on religion—provided it is fully voluntary and does not result from or lead to injustice" (IPRR, 161). Role inequalities might be considered compatible with a political conception of self-respect as a free and equal citizen. But the point, again, is that self-respect develops in childhood, and that children are not capable of making this distinction. If the gendered division of labor or a religious teaching deprives children of self-respect, it has led to an injustice.

The question, then, is how a liberal state can best address the expression of negative attitudes within the family and their effects on children's self-respect. Restricting parental speech appears to interfere with parental rights. In the case of playground name-calling, by contrast, the intervention of a teacher or aide poses no such problem of interference: not only are children fitly subject to paternalistic intervention by legitimate authorities such as teachers, but teachers and schools have a right (and probably a duty) to intervene in expressions of incivility. What might seem more problematic, though, is the attempt to stop the proliferation of negative stereotypes by interfering with parental "infusion" of such attitudes in children. Where parents have religious freedom and a strong interest in the parent-child relationship, they might be thought to have a right to infuse their child with their beliefs.

Educational attempts to address sexism, racism, and homophobia directly, as in civics education, and indirectly, through the selection of textbook examples or works of literature, are sometimes met with this protest. Where oppression on the basis of sex, race, or sexual orientation threatens children's self-respect, such prophylactic education is an important means of securing self-respect (as well as of creating future citizens of a liberal state). But such classroom measures as depicting same-sex couples positively create controversy based on parents' alleged rights to infuse their child with their beliefs. What should a liberal state do when faced with parents' claim to a right to infuse their children with racist, sexist, or homophobic beliefs and attitudes that stand to undermine self-respect by inculcating beliefs that some kinds of people have inferior status and worth?

The teaching of such beliefs may have negative effects on the self-respect of the child infused as well as on other children consequently subjected to playground bullying and stigmatization. Consider the case of religiously motivated parents infusing children with "deferential wife" gender roles and homophobic attitudes. Such infusion will undermine recognition self-respect if it transmits beliefs that certain kinds of people have less value than others, that their plans have less worth, and that their agency has less force. For example, children might be taught that wives' preferences matter less than husbands'. This seems reasonably likely to undermine the self-respect of girls so taught, and also of girls from other families with whom these girls interact.

This might seem to suggest a conflict between the social bases of self-respect—children's freedom from stereotypes that undermine their self-respect—and parental rights that fall under the category of religious freedom. Against the claim that gender roles are protected by religious freedom, Okin stressed the importance for feminism of protecting girls' self-respect (Okin 1994; 2005, 242).[13] But, as noted above, Rawls suggests that some gendered division of labor within the family is protected by religious freedom. If freedom of religion is itself a social basis of self-respect, as Rawls's argument for the priority of the liberty principle implies, then there appears to be a conflict between two bases of self-respect.

But the conflict is merely apparent: parental rights to infuse children should be distinguished from religious liberties. The latter are self-regarding; religious liberty does not give one the right to compel others to adopt one's religion. Yet this is what infusion amounts to: children generally cannot choose whether to be infused with such beliefs. Their participation is not fully voluntary, as Matthew Clayton has insisted (2006). In benign cases, it is reasonable to treat such infusion on a par with other sorts of infusions—of tastes in food or art or sports. The preferences that children learn from parents are not acquired fully voluntarily—but this does not generally undermine the validity of such preferences. Religious affiliation is considered voluntary from a political point of view (IPRR, 161–62; see also Levey 2005). But when such infusion threatens the self-respect of the child involved, then the parent's pursuit of his conception of the good directly, perhaps permanently, threatens a primary good for another, nonconsenting, individual. Depriving someone of a primary good of equal status with the liberties, without her consent, threatens to pass from speech to a violation of rights; thus Rawls writes that the tolerant must uphold the free speech of the intolerant unless there are "considerable risks to our own legitimate interests," such as liberty (TJ, 219).

Parental rights within Rawlsian liberalism should be treated like other entitlements—as "derived from social institutions [structured by the principles of justice] and the legitimate expectations to which they [the institutions] give rise" (*TJ*, 10). Plausible accounts of parental rights, such as the status of parenting as a unique and irreplaceable project (Brighouse and Swift 2006), do not justify jeopardizing the child's self-respect or that of other children with whom the child interacts. We should understand parental rights as constructed in view of the principles of justice. And this means that the legal protection of children's developing self-respect, as a right that is a social basis of self-respect, should be secured within legal parenting frameworks, just as those frameworks prohibit abuse and neglect.

The status of self-respect as an important primary good limits parents' rights to infuse children with their beliefs or to subject them to systematic and arbitrary unequal treatment. In practice, it might be best to encourage such limits through parent education and mandatory school curricula designed to reinforce self-respect and counteract teachings that undermine self-respect. Owing to the difficulty of intervening in parent-child communication in isolated nuclear families, policies that directly restrict parental speech seem unlikely to be effective (although teachers might look for signs of damaged self-respect just as they look for signs of physical abuse). Because young children need continuity of care for healthy cognitive development (Alstott 2004), and because older children's self-respect may be bound up with family pride, there are strong child-welfare reasons not to remove children from parental custody in most cases. But this approach leaves children open to a kind of "schizophrenia," whereby parental teachings clash with what children are taught in school.[14] To avoid placing children in this position in the first place, the status of self-respect as a primary good might justify parental licensing, as proposed by Hugh LaFollette (1980)—although this poses other challenges. At least the removal of children from parental custody when their developing self-respect is threatened with serious and irreversible damage might be justified. While the best way to protect children's developing self-respect is unclear, the theoretical point should be clear: parental rights do not entail a right permanently to deprive children of an important primary good.

I have argued that self-respect, understood as recognition self-respect, is a Rawlsian primary good, underwriting citizens' pursuit of plans in various ways. Because of the special vulnerability of children's self-respect, and the effects of damage to their self-respect on their life chances, treatment and teachings that threaten their self-respect constitute an injustice that the theory should not tolerate. This in turn provides a Rawlsian strategy for feminist criticism of the construction of gender roles within the family, even if those roles do not lead to economic inequalities. It suggests a principled limit on political liberalism's toleration of inequality within family life.

Notes

1. Arneson 1990 charges primary goods with insensitivity to how citizens actually fare; people can have many primary goods yet little welfare. But subjective accounts also face problems, such as drainage of resources by the neediest.

2. Such neutrality, though controversial, and restricted in scope in *PL*, is a hallmark of Rawlsian liberalism.

3. Thus there is a nonneutrality at the base of Rawls's theory—a claim that the moral powers are the basis of moral equality and should define the political understanding of the person.

4. This does not fit neatly with Rawls's comments on the Aristotelian principle in *TJ* (440–42); because he conflates appraisal and recognition self-respect, his comments cannot all be worked into one coherent interpretation.

5. Rawls addresses a similar objection to his claim that associations will support self-respect, asking whether this applies only to elite associations (*TJ*, 441). The fact that he does not there address the otherwise glaring problem developed here suggests that his immediately preceding argument for self-respect does not involve empirical self-appraisal.

6. Thus I agree with Doppelt 2009, which argues that Rawls should have endorsed recognition, not appraisal, self-respect—although I am less convinced than Doppelt is that Rawls had only appraisal self-respect in mind.

7. Self-respect faces the objection that primary goods are not truly neutral: any good—money, self-respect, liberties—may be useless, or less useful, for some plans of life, e.g., money for mendicant monks or self-respect for someone practicing religious self-mortification (Nagel 1973; Schwartz 1973). In response, primary goods are normally (not universally) needed for the pursuit of plans of life (prompting the difficult question of what "normally" means here) (*PL*, 76).

8. This may leave a gap in Rawls's theory of justice, however, if the argument for that theory depends on the primacy of self-respect, which provides a crucial argument for the priority of liberty and hence the lexical ordering of the principles. See Eyal 2005, 198–99.

9. Nussbaum's discussion in chapter 7 of *Sex and Social Justice* (1999) is illuminating on the issue of how such categorization masks the subtleties and complexity of sexual desire and behavior—for instance, men in all-male milieux (e.g., prisons) who practice same-sex activities are often not defined, nor do they see themselves, as "homosexuals."

10. See Calhoun 2003. In the United States, the promotion of marriage has been mandated since 1996 under the Personal Responsibility and Work Opportunity Reconciliation Act, through programs such as abstinence-only education and the Healthy Marriage Initiative.

11. This study measured self-esteem on the basis of children's responses to statements like "I take a positive attitude toward myself," which reasonably relate to recognition self-respect, as well as on the basis of emotional and behavioral symptoms.

12. Rawls alludes to this passage in Rawls 2008, 299.

13. The role of religion in children's developing self-respect deserves further attention. Some studies have found that religious belief improves self-image, but this effect is stronger among boys than girls (Lewis and Pearce 2008).

14. Thanks to David Archard for drawing my attention to this point; see Okin 1994 and Abbey 2011, 74–75.

4

"The Family as a Basic Institution"

A Feminist Analysis of the Basic Structure as Subject

Clare Chambers

In Section 50 of *JFR*, titled "The Family as a Basic Institution," Rawls replies to Okin's feminist critique of *TJ*. He states, "If we say the gender system includes whatever social arrangements adversely affect the equal basic liberties and opportunities of women, as well of those of their children as future citizens, then surely that system is subject to critique by the principles of justice" (*JFR*, 167–68). As the introduction to this volume demonstrates, the question of how Rawlsian justice might secure gender equality has been discussed by many feminists, most notably Okin. However, the Rawls-Okin debate raises more questions than it answers. Okin

---

I thank Ruth Abbey, Chris Brooke, Andy Mason, Miriam Ronzoni, Peter Stone, and Andrew Williams for written comments on earlier versions of this chapter.

criticizes Rawls for failing to apply his theory adequately to the family: she criticizes not Rawls's approach in general but his attitude toward the family in particular. Okin argues that a consistent application of Rawlsian theory would secure gender justice but that Rawls is remiss in refusing such consistency. In fact, as I show, Rawls's remarks on the family reveal a more fundamental problem with Rawlsian theory. It is not that Rawls fails to apply his theory correctly to the family, but rather that the specific case of the family illustrates deep-seated difficulties with Rawlsian justice as a whole.

The problem is that Rawls's ambiguous remarks on the family are comprehensible only at the expense of his fundamental claim that there is something distinctive about the application of justice to the basic structure. Okin criticizes Rawls for failing to make good on the fact that the family is part of the basic structure. If he did make good, Okin claims, he would see that the principles of justice must apply to the family in a much more extensive way than he actually allows. As I show, however, the family is one illustration of the fact that how the principles of justice apply to an institution does not depend on whether that institution is part of the basic structure. This is a problem for Rawls because the distinctiveness of the basic structure is a crucial part of the political liberalism that, by the end of his work, has become essential to the Rawlsian project.[1]

I first outline Okin's critique of Rawls and provide a valid formalization of her argument. I then examine the main premises of her argument and look for evidence to support Okin's interpretation of Rawls. I conclude that it is flawed but nonetheless highlights problems with Rawls's claim that the basic structure is the subject of justice. I consider and reject the argument that Rawls's theory is consistent according to what I call the "whole structure view": that the principles of justice apply to the basic structure considered as a whole. Finally, I consider G. A. Cohen's argument that the basic-structure distinction is problematic. I agree with this criticism, but I suggest that Cohen is wrong in situating the problem with the issue of coercion. I conclude that Rawls's position on justice in the family is at odds with his claim that the basic structure is uniquely the subject of justice.

Okin's Critique

Okin argues in JGF that Rawlsian justice has the potential to secure gender equality but that Rawls fails to bring out this potential. Rawls fails to note three consequences of his stipulation that sex is one of the unknown characteristics behind the veil of ignorance. First, that stipulation seriously undermines Rawls's claim that the parties in the OP are heads of households, since that implies that they are male and thus perpetuates patriarchal divisions of labor. Second, if sex were behind the veil of ignorance, then those in the OP would be greatly concerned about matters that Rawls does not discuss, such as "many aspects of social gendering and sex discrimination as well as matters affected by biological sex differences" (Okin 2005, 237–38). Third, if sex were unknown, then "families would certainly have to be taken seriously as part of the basic structure of society" (238).

Okin is right to press the first two problems, and I do not discuss them further.[2] The third problem is the focus of this chapter: that the family needs special consideration because it is part of the basic structure. Okin argues that *PL* makes the problem worse, because in it and later works Rawls denies that the principles of justice apply to the family. Three passages from Okin illustrate her critique:

> Though he lists "the nature of the family" as part of the basic structure, in *Political Liberalism* he also explicitly places families on the non-political side of the public-private divide. . . . Rawls also asserts that the family is like other "voluntary institutions," such as churches and universities, firms and labor unions, and thus is not itself subject to the principles of justice or expected to run democratically. But this notion is simply *implausible*. . . . The notion that families are not distinct from other voluntary institutions is also completely inconsistent with the crucial place that supposedly just families play in Rawls's theory of early moral development. (2005, 241)

> If the family, unlike the other voluntary associations, is both *part of the basic structure of society and the place where a sense of justice is first developed in the young*, then does it not need to be internally just? (2005, 245)

> Only by allowing that his principles of justice apply directly to the internal life of families—which Rawls clearly resists . . . could one revise the theory so that it both includes women and has an effective and *consistent* account of moral development. (2004, 1538–39, emphasis added)

Thus Okin argues that Rawls fails to make good on the implications of his theory: that, as part of the basic structure, the principles of justice should apply internally to the family. Indeed, he explicitly denies that they should do so, drawing an "implausible" analogy between the family and voluntary associations. The result is that Rawls's theory "contains an internal paradox" (JGF, 108).

The idea that the family should be "internally just," together with Rawls's formulation that discusses whether principles of justice should apply "to the internal life" of the family, is not fully specified. I discuss this terminology in detail below. For now, the family will be considered internally just if, considered in isolation, activities and distributions within it conform to the principles of justice. So family members must enjoy equal basic liberties and equal opportunities with respect to family concerns, and any inequalities within the family must benefit the worst-off members.

The final phrase of this sentence is deliberately ambiguous: if the family is to be internally just, should inequalities benefit the worst-off members of the family or the worst-off members of society? Okin certainly claims at least the latter: Rawls is wrong, she argues, to consider heads of households as the units of distribution for the difference principle, and so we must look inside the family to consider the distributive shares of individual members when assessing whether a society conforms to the difference principle. This understanding of the family being internally just is one that I think Rawls both should and could accept. However, Okin also defends the former interpretation: that there should be a separate difference principle governing inequalities within the family, viewed in isolation from society as a whole.[3] I discuss and criticize this interpretation later in the chapter.

A valid formalization of Okin's argument has the following form:

1. The family is part of the basic structure.
2. Rawls's theory entails that the principles of justice should apply internally to institutions that are part of the basic structure (but not to others).
3. Therefore, for Rawls, the principles of justice should apply internally to the family.
4. Rawls states that the principles of justice should not apply internally to the family.
5. Therefore, Rawls's position on the family is inconsistent.

This argument is valid, but it is not sound. Rawls does consistently affirm premise 4. For example, he states, "We wouldn't want political principles of justice—including principles of distributive justice—to apply directly to the internal life of the family" (IPRR, 159; see also JFR, 163 and 165). Rawls also repeatedly affirms premise 1; I discuss some ambiguities of that premise below but conclude that they do not undermine Okin's critique. As for premise 2, however, Rawls does state the claim in brackets but does not affirm the premise as a whole, and his theory does not entail that he should affirm it. Instead, Rawls argues that the principles should apply directly to basic-structure institutions, and this locution does not carry the same implications as the idea of internal application.

I expand and reference these claims in the next sections, and show why Okin's argument that the principles of justice should apply internally to the family rests on a misunderstanding of Rawls. It is not new to say that Okin's interpretation of Rawls is problematic.[4] But my claim is that Okin's critique of Rawls nevertheless highlights very deep problems for Rawls's theory, problems that extend beyond family justice. The reason that Okin's interpretation is wrong is that Rawls cannot (indeed, does not) support his central claim that the principles of justice apply directly to the basic structure but not to other institutions. Principles of justice do apply directly to some institutions but not to others, but the difference is not explained by the basic-structure/non-basic-structure distinction. Okin is wrong about what Rawls claims for his own theory, then, but Rawls's theory is left extremely muddled. While the argument is invalid, it remains possible that both conclusions (3) and (5) are true.

Okin's Argument in More Detail

One passage from JFR provides some support for Okin's interpretation of Rawls. The bracketed numerical insertions indicate support for the relevant premises of the formalization of Okin above.

> [1] The family is part of the basic structure, the reason being that one of its essential roles is to establish the orderly production and reproduction of society and of its culture from one generation to the next. . . . The primary subject of justice is the basic structure of society understood as the arrangement of society's main institutions into a unified system of social cooperation over time. [2] The principles of political justice are to apply directly to this structure, but they are not to apply directly to the internal life of the many associations within it, [4] the family among them. (JFR, 162–63)[5]

Insofar as this supports Okin's argument, it reveals Rawls making what seem straightforwardly contradictory statements within the space of two short pages. For Rawls claims that the family is part of the basic structure, and that the principles of justice should apply directly to the basic structure, but that they should not apply directly to the family. The extent of this apparent contradiction gives pause for thought. Rawls is a serious, enormously important political philosopher, and it seems incredible that his view could be quite so muddled. How, then, might we reconcile Rawls's apparent contradiction? I do not attempt any reinterpretation of premise 4: wherever he makes remarks on the subject, Rawls does indeed state that the principles of justice should not apply internally to the family. Instead, I highlight ambiguities that modify premises 1 and 2.

Premise 1: The Family Is Part of the Basic Structure

There are many places in which Rawls claims unambiguously that the family is part of the basic structure. As well as in the excerpt from *JFR* quoted above, we find this claim made right from the start of *TJ*: "For us the primary subject of justice is the basic structure of society, or more exactly, the way in which the major social institutions distribute fundamental rights and duties and determine the division of advantages from social cooperation. . . . Thus the legal protection of freedom of thought and liberty of conscience, competitive markets, private property in the means of production, and the monogamous family are examples of major social institutions" (7). The family is part of the basic structure, then, because the basic structure is composed of the major social institutions, and the family is an example of a major social institution.

Perhaps, however, there are two ways that one thing can be part of another. Imagine that you and I stand on a driveway and I draw a chalk circle around us. We are each "part of" the circle in the sense that we stand within it and are encompassed by its boundaries. But we are not "part of" the circle in the sense that you and I are not particles of the chalk that makes up the circle. Some distinction of this sort may be going on in Rawls. He characterizes churches and universities as associations that are "within" the basic structure, and this category of being "within" seems distinct from the category of actually being part of the basic structure: "Since justice as fairness starts with the special case of the basic structure, its principles regulate this structure and do not apply directly to or regulate internally institutions and associations *within* society" (*JFR*, 10, emphasis added). "*The first principles of justice as fairness are plainly not suitable for a general theory.* . . . The most we can say is this: because churches and universities are associations *within* the basic structure, they must adjust to the requirements that this structure imposes in order to establish background justice" (PL, 261, emphasis added). Although churches and universities are "within" the basic structure, they are not themselves part of the basic structure: Rawls talks about the structure imposing requirements on them, and it could not do this if the two were equivalent. This passage, then, suggests a hierarchy between institutions that make up, that are, the basic structure, and associations within it.[6] Churches and universities stand within the circle of the basic structure, but they are not themselves made of chalk.[7]

This distinction would solve the contradiction about the family if Rawls's statement "the family is part of the basic structure" (JFR, 162) meant "the family is within the basic structure," not "the family (partially) constitutes the basic structure." Such an interpretation would make sense of his otherwise puzzling claim that "the principles of political justice are to apply directly to this [basic] structure, but they are not to apply directly to the internal life of the many associations within it, the family among them" (JFR, 163; cf. 10). On this interpretation, there would be no inconsistency in Rawls, and Okin's critique would be wrong. For when she says that the principles of justice should apply internally to the family because the family is part of the basic structure, she means that they should so apply because the family constitutes the basic structure. But the family would not (on the hypothesis currently under consideration) constitute the basic structure, and so there would be no reason why the principles of justice should apply directly to it.

There are several problems with this solution, however. First, it is not the usual interpretation of Rawls. Of course, this in itself is not conclusive: the usual interpretation could be wrong. But one might think that Rawls would have attempted to correct the usual interpretation if it were wrong. Not only did he not correct it, but the words he uses invite the usual interpretation. The phrase "the family is part of the basic structure" suggests more strongly that the family makes up the basic structure than that it is merely an association within the basic structure. After all, if churches and universities are associations "within" the basic structure, then it seems that all associations are similarly placed. Being "within" the basic structure looks like simply being "within" society. There is no need to single out the family with statements like "the family is part of the basic structure" if all that is meant is "the family is an association within the basic structure, like every other association that exists within a society." Moreover, Rawls refers elsewhere to the family as an "institution of the basic structure" (IPRR,

163), a phrasing that once again suggests that the family makes up the basic structure.

There are also good reasons within Rawls's theory to think that the family is part of the basic structure in the sense of constituting it, reasons that Okin rightly highlights. Rawls argues that the family does have a special place within society: it is a basic institution in that it plays an essential role in reproducing society and ensuring that citizens have the sense of justice that is necessary for society to be well ordered. The family is different from churches and universities because of these extra features (Freeman 2007, 101–2). Indeed, Rawls makes these claims as an explanation of the claim that the family is part of the basic structure (*JFR*, 162).

There thus remains an apparent contradiction between Rawls's claim that "the family is part of the basic structure" to which principles of justice are to "apply directly," and the claim that the principles of justice "are not to apply directly to the internal life of the many associations within it, the family among them." This contradiction cannot be removed by modifying premise 1: the family is a central part of the basic structure, and if there is a distinction between institutions that make up the basic structure and those that are merely within it, the family belongs in the former category.

I turn now to premise 2 of the formalized version of Okin's critique: Rawls's theory entails that the principles of justice should apply internally to basic-structure institutions (but not to others). Rawls certainly states that the principles of justice should not apply to the internal life of non-basic-structure institutions: he affirms the bracketed part of premise 2. In *JFR* he writes, "Since justice as fairness starts with the special case of the basic structure, its principles regulate this structure and do not apply directly to or regulate internally institutions and associations within society" (10).

Notice, though, that in this formulation there are two things that the principles of justice do not do to nonbasic institutions: they neither "apply directly" to nor "regulate internally" such institutions. This ambiguity is essential. The natural implication of many of Rawls's statements is that the principles of justice do "apply directly to" and "regulate internally" the institutions of the basic structure. However, while Rawls makes many references to the principles' applying directly to the basic structure, he makes no reference that I can find to the principles' regulating basic-structure institutions internally. Whenever he explains how the principles of justice differ with regard to the basic and nonbasic institutions, he uses the idea of internal application only to describe what does not happen for institutions that are not part of the basic structure. The excerpt quoted above repeats this formulation: "The principles of political justice *are to apply directly* to this [basic] structure, but they are *not to apply directly to the internal life* of the many associations within it" (*JFR*, 163, emphasis added).

If there is a difference between the principles' applying "directly" and applying "internally" that is to rescue Rawls from Okin's critique of inconsistency, then it must be the case that the principles of justice do not apply "directly to the internal life" of any institutions, whether part of the basic structure or not. Since we are told explicitly both that the family is part of the basic structure and that the principles do not apply directly to its internal life, it must be that the principles of justice simply do not apply to the internal life of basic-structure institutions (cf. Lloyd 1995, 1326–27).

Moreover, as we have seen, Rawls explains that the principles do not apply to the internal life of non-basic-structure institutions such as churches and universities. So Rawls introduces a red herring in the many passages where he discusses the basic-structure/non-basic-structure distinction alongside the lack of internal application of the principles: the two concepts cannot be related if he is to remain consistent. The principles of justice do not apply to the internal life of any institutions. However, the analysis cannot stop here, because a central aspect of Rawls's theory is the idea that justice applies in a special way to the basic structure. There must therefore be something that the principles do to the family "as part of the basic structure." The alternative to the idea of internal application that we find in Rawls is the idea of "direct" application.

There are several passages that both (1) support the Rawls-rescuing idea that there is a crucial difference between direct application and internal application and (2) demonstrate how Rawls introduces the red herring of discussing claim (1) alongside references to the basic structure. For example, he writes,

> Political principles do not apply directly to [the family's] internal life but they do impose essential constraints on the family as an institution and guarantee the basic rights and liberties and fair opportunities of all its members. This they do . . . by specifying the basic claims of equal citizens who are members of families. The family as part of the basic structure cannot violate these freedoms. Since wives are equally citizens with their husbands, they have all the same basic rights and liberties and fair opportunities as their husbands. (*JFR*, 164)

So the principles of justice do apply to the family in the sense that the family may not violate the freedom and equality of its members as citizens. Husbands cannot prevent their wives from voting, for example—one of the equal basic liberties that all citizens enjoy—and wifehood cannot be a status that legally removes one's right to vote. However, this requirement of justice does not apply to the "internal life" of the family. Wives are not required to have a vote inside the family on matters of domestic life.

Can we say that we have found the way in which the principles of justice apply "directly" to the family? Remember that in order to support Rawls's basic-structure/non-basic-structure distinction, we are looking for a way in which the principles do apply to the basic structure but not to other institutions (such as churches and universities). This idea that the principles apply directly by imposing "essential constraints" on basic institutions is supported by Rawls's claim, quoted just above, that the family, "as part of the basic structure," may not violate the "basic rights and liberties and fair opportunities of all its members." But there is a problem: the phrase "as part of the basic structure" is unnecessary, since no associations, whether part of the basic structure or not, may violate the equal basic liberties of their members as citizens. A husband may not remove his wife's legal right to vote, but then neither may a church remove that right from its members, nor a university from its employees and students. It is not because the family is part of the basic structure that it cannot interfere with the rights of citizens. As Rawls himself puts it, "Even if the basic structure alone is the primary subject of justice, principles of justice still put essential restrictions on the family *and all other associations*. The adult members of families and other associations are equal citizens first: that is their basic position. *No institution or association in which they are involved can violate their rights as citizens*" (JFR, 166, emphasis added).[8] So imposing "essential constraints" or "essential restrictions" cannot be the thing that the principles of justice do uniquely to the basic structure, and cannot be what is meant by the principles' applying "directly."

This and similar passages undermine the idea that whether the principles of justice apply to an institution depends on whether that institution is part of the basic structure. There seems to be no difference between the way the principles of justice apply to the family "as part of the basic structure" and the way they apply to it simply as an association like any other. If it is correct to say that the requirement not to violate people's rights and liberties as citizens means that the principles of justice apply "directly" to the family, then it would also be true to say that the principles apply directly to institutions that are merely within, and do not constitute, the basic structure. Rawls does not explain in what special, additional ways the principles of justice apply to the family as a basic institution.

The Basic Structure as Subject

This problem is of wider application. A fundamental aspect of Rawls's theory is the idea that the principles of justice apply differently to basic-structure than to non-basic-structure institutions. Consider, for example, his remarks in *JFR*: "One main feature of justice as fairness is that it takes the basic structure as the primary subject of political justice. . . . One should not assume in advance that principles that are reasonable and just for the basic structure are also reasonable and just for institutions, associations, and social practices generally" (10–11). The distinctiveness of the basic structure is a "main feature of justice as fairness," because it is that distinction that preserves the political (i.e., noncomprehensive) nature of Rawls's liberalism. By limiting justice to the basic structure, Rawls is able to put forward a political conception of justice that stands apart from any comprehensive conception of the good: it neither relies on a conception of the good for its support nor imposes standards that adherents of a reasonable conception of the good could not accept. Without the basic-structure distinction, Rawlsian liberalism loses this distinctive feature.[9]

The mechanism for the supposedly freedom-enhancing politically liberal distinction between the basic structure and all else lies in a difference in the way that justice applies to the basic structure and the way it applies elsewhere: Rawls describes an "institutional division of labor between the basic structure and the rules applying directly to individuals and associations and to be followed by them in particular transactions" (*PL*, 268–69). This is not to say that justice does not apply at all to institutions that are not part of the basic structure: "there is no such thing" as "a space exempt from justice" (IPRR, 161). Rawls is not, therefore, employing a traditional public-private distinction. But there are certainly supposed to be differences between the application of justice in the basic structure and its application elsewhere.

Rawls does provide an account of what it means to say that the principles of justice do not apply internally to the family, but it is not enormously helpful. "We wouldn't want political principles of justice to apply directly to the internal life of the family," he writes. "It is hardly sensible that as parents we be required to treat our children in accordance with political principles. Here those principles are out of place. Certainly parents should follow some conception of justice (or fairness) and due respect in regard to each of their children, but, within certain limits, this is not for political principles to prescribe" (*JFR*, 165). There are various problems with this passage. First, Rawls appeals to the intuition that we would not want political principles to apply to the relationships between parents and children. But this intuition is not explained by the basic-structure/internal-life distinction, since in no area of society are children accorded equal basic liberties. Parents should not treat their children according to political principles because children should not be treated according to such principles, not because such treatment occurs "inside" the family.

A more pertinent question, then, is whether adult members of a family should relate to one another according to the political principles of justice. Rawls discusses this question in only one work, IPRR, where he writes that "the government would appear to have no interest in the particular form of family life, or of relations among the sexes, except insofar as that form or those relations in some way affect the orderly reproduction of society over time. . . . Of course, there may be other political values in the light of which such a specification would pass muster: for example, if monogamy were necessary for the equality of women, or same-sex marriages destructive to the raising and educating of children" (147). These remarks have the potential for a fairly radical feminist interpretation, particularly given Rawls's aside that "I should like to think that Okin is right" (JFR, 167). After all, Okin claims that the family's role as a school of justice requires an equal division of labor between women and men, and if Rawls accepts that the "equality of women" is a political value that appropriately constrains the family, then the potential for feminist reform of the family might be considerable. This "Eureka!" moment is short-lived, however, for Rawls once again introduces caveats that both limit the egalitarian potential of his theory and further undermine the distinctiveness of the basic structure. IPRR continues:

> Some want a society in which division of labor by gender is reduced to a minimum. But for political liberalism, this cannot mean that such division is forbidden. One cannot propose that equal division of labor in the family be simply mandated, or its absence in some way penalized at law for those who do not adopt it. This is ruled out because the division of labor in question is connected with basic liberties, including the freedom of religion. Thus, to try to minimize gendered division of labor means, in political liberalism, to try to reach a social condition in which the remaining division of labor is voluntary. This allows in principle that considerable gendered division of labor may persist. It is only involuntary division of labor that is to be reduced to zero. (162)

There is a great deal of literature criticizing the concept of voluntariness as it applies to gender inequality (see Chambers 2008). The key point for our purposes is that Rawls's remarks could apply equally to non-basic-structure institutions, such as churches. Much of the literature concerning political liberalism and Rawlsian multiculturalism centers on Rawls's claim that while cultures and religions must refrain from the involuntary imposition of inequality, there can be no political critique of or coercive intervention in voluntary inequality (Chambers 2008). So, once again, there is no uniqueness in the application of the principles of justice to the basic structure.

The Whole-Structure View

I began this chapter by considering Okin's question to Rawls: if the family is part of the basic structure, and if the principles of justice apply directly to the basic structure, shouldn't the principles of justice apply to the internal life of the family? We have seen that Rawls's insistence that the principles of justice do not apply to the internal life of the family must rely on the premise that they do not apply to the internal life of any institution. Applying to the internal life of the relevant institutions cannot be what Rawls means when he says the principles of justice apply directly to the basic structure.

What, then, does Rawls mean by saying that the principles of justice apply directly to the basic structure? In this section, I consider an answer that Rawls might give, an answer that I call the whole-structure view. Several commentators argue that the whole-structure view is the right way to understand Rawls.[10] The whole-structure view emphasizes the structure part of the basic structure, and states that the principles of justice apply directly to the basic structure of society in the sense that that they apply directly to the ways in which the major institutions interact with one another to form a whole. Justice must be considered as a virtue of this overall structure, not as a virtue of each institution considered in isolation. The principles do not apply to the internal life of the family (or any other institution), because that is only one institution among the many that together form the structure. The site of justice is the interactive whole, not the isolated component parts.

Amy Baehr argues that the whole-structure view rescues Rawls from Okin. Whereas Okin "would have the family be directly subject to the two principles of justice," in fact "the principles of justice are not to be applied directly to any one institution but to the basic structure as a whole" (Baehr

1996, 51). It follows, Baehr argues, that the family itself might be unproblematically unjust, since its injustice might be compensated for elsewhere: "Rawls . . . points out that the system taken as a whole may be just even where some single institution is unjust (*TJ*, 57). This may be the case when relations in one institution make up for injustice in another. The claim of justice as fairness is that the system of institutions as a whole must be just" (52). The example that Baehr gives is the difference principle. The difference principle does not apply internally to the family, in the sense that the family does not have to distribute its resources so that any inequality benefits the worst-off family member. Baehr argues, contra Okin, that this is not because Rawls inconsistently forgets that the family is part of the basic structure, for the difference principle does not apply internally to any institution. There is one difference principle for the whole of society: inequalities are permitted if they benefit the least advantaged, considered within society as a whole.[11]

The whole-structure view is sometimes introduced to show that Okin wrongly interprets Rawls (see Abbey 2011, 62–65). As a separate question, we may ask: if Rawls does in fact hold the whole-structure view, does this secure the distinctiveness of the basic structure as a unique site for the application of the principles of justice? In order for the whole-structure view to rescue the distinctiveness of the basic structure, two further claims must hold. First, it must in general be true that justice is concerned with the structure as a whole, not with any individual part of it; in other words, this claim must hold for each of the principles of justice. Second, it must also be true that only those institutions that are part of the basic-structure institutions are relevant for the overall justice of society. Without this second claim, there is no role for the basic structure: justice would be a feature of a society *tout court*, and the concept of the basic structure would be redundant.

Neither claim stands. First, it is not true in general that each of the Rawlsian principles of justice applies only to the basic structure as a whole and not to its various individual institutions. Baehr is correct in arguing that a whole-structure view must be taken for the difference principle, and that is important. But of all the principles of justice, the difference principle has the lowest priority and is not a constitutional essential. The two more important principles are not secured by taking an overview approach that ignores the justice of individual institutions. Second, these two principles (the equal-basic-liberty and the equal-opportunity principle) must apply directly not only to the individual institutions of the basic structure but also to institutions that do not constitute it.

We have already seen that justice cannot be achieved if some institutions deny the basic liberties, so all institutions are subject to the constraints of the first principle (cf. Abbey 2007, 14). A husband cannot prevent his wife from voting in the sense of imposing legal limits on her right. And these restrictions apply to all institutions, whether basic or not. Rawls states that "the principles defining the equal basic liberties and opportunities of citizens always hold in and through all so-called domains. The equal rights of women and the basic rights of their children as future citizens are inalienable and protect them wherever they are" (IPRR, 161).

The equal-basic-liberty principle thus disproves each of the two claims essential to the whole-structure view. Moreover, each claim is also undermined by the equal-opportunity principle. The excerpt from Rawls just quoted also refers to the equal-opportunity principle: this too cannot be realized for the basic structure overall if it is disregarded by some institutions. Imagine that one basic-structure institution, such as the department of health, refuses to hire women, while another, such as the department of education, refuses to hire men. Even if the two departments have the same number of equivalent jobs, this is not a society characterized by equal opportunity—just as it is not if some state schools accept only white pupils while others accept only black pupils. Rawls appears to recognize this fact when he states that the equal-opportunity principle governs "all sectors of society" (TJ, 73), and indeed it could not be otherwise: if only some positions are allocated according to equal opportunity, then it follows that there is not equal opportunity overall.

Moreover, Rawls agrees that non-basic-structure institutions are also properly restricted by the demands of equal opportunity (PL, 261). A society does not have equal opportunity if government departments hire according to sexist criteria, but it also lacks equal opportunity if private employers do the same. Indeed, the equal-opportunity principle applies to institutions of the basic structure in just the same way as it applies to other institutions, posing further problems for the distinction between the principles applying "directly" and those applying "internally." Equal opportunity cannot exist in a society unless it applies to internal decisions made by institutions: most notably, employment and promotion decisions of employing organizations. As I have argued elsewhere (Chambers 2009), equal opportunity cannot be achieved or even respected merely by institutions that fail to violate the equal opportunity that is somehow brought about elsewhere. Employers must make hiring decisions in accordance with the demands of equal opportunity (just as current antidiscrimination law requires that they must make hiring decisions without reference to certain characteristics such as race and sex). And although the economic order is part of the basic structure, Rawls suggests that organizations such as "business firms" within it are not.[12] So the equal-opportunity principle must apply directly to the internal life of (some) non-basic-structure institutions.

Rawls is not rescued, then, by the whole-structure view. Rawlsian justice cannot be secured unless individual institutions are just, including the family. But we have not found a way in which membership in the basic structure determines the application of the principles of justice.

The Basic Structure and Coercion

My argument so far is twofold. First, I have argued that Okin is wrong to claim that Rawls's own theory requires that the principles of justice apply to the internal life of the family. Although Rawls does say that the principles of justice apply uniquely to the basic structure, he does not say that they apply to the internal life of the institutions of the basic structure. Second, I have argued that Okin's critique highlights a more general problem with Rawls's account: he does not, and possibly cannot, justify his claim about the specialness of the basic structure.

G. A. Cohen makes an argument that has similarities to mine. Cohen argues, correctly, that it is "seriously unclear which institutions are supposed to qualify as part of the basic structure" (2008, 132). He locates this ambiguity in the question of whether the basic structure is composed only of legally coercive institutions (or the legally coercive aspects of institutions), and rightly argues that it is wrong to care about coercive structures and not also about "the ethos that sustains gender inequality and inegalitarian incentives" (138). Both, as Cohen rightly insists, are crucial to explaining the existence of injustice and to rectifying it. Cohen sees this fact as utterly undermining Rawls's insistence that the principles of justice apply uniquely to the basic structure. I agree with Cohen's conclusion that Rawls's statements about the uniqueness of the basic structure are problematic. I also agree that justice must be concerned with noncoercive phenomena as well as with what is coercive (Chambers 2008). However, I do not agree with Cohen that the distinction between coercive and noncoercive institutions is the source of the instability in Rawls's account.

Cohen argues that Rawls faces a dilemma. Rawls could say that the basic structure is limited to the coercive aspects of institutions, but then he would face the objection that noncoercive aspects of society can also have the profound effects that justify restricting justice to the basic structure in the first place (Cohen 2008, 136). Or Rawls could admit that justice is affected by noncoercive phenomena (such as whether citizens have an egalitarian ethos), but then the idea that justice applies uniquely to the basic structure collapses (137).[13] Cohen illustrates his position with the example of the family and the sexist expectations that may attach to the roles of husband and wife:

> Such expectations need not be supported by the law for them to possess informal coercive force: sexist family structure is consistent with sex-neutral family law. . . . Yet Rawls must say, on pain of giving up the basic structure objection, that (legally uncoerced) family structure and behavior have no implications for justice in the sense of "justice" in which the basic structure has implications for justice, since they are not a consequence of the formal coercive order. But that implication of the stated position is perfectly incredible: no such differentiating sense is available. (137)

I agree with the tenor of Cohen's remarks. But I do not think that they are as crushing for Rawls as Cohen thinks, for two reasons. The first is that, contra Cohen, Rawls does have a "differentiating sense" available to distinguish the justice of the basic structure from justice in other spheres. In his later work he distinguishes three kinds of justice: local, domestic, and global. Rawls would characterize the distinction that Cohen thinks is impossible as a distinction between "local justice (principles applying directly to institutions and associations)" and "domestic justice (principles applying to the basic structure of society)" (JFR, 11). Whether this distinction is a good one is difficult to say, since Rawls refuses to discuss local justice any further. But Cohen does not provide sufficient argument to show that the justice of the basic structure must be exactly equivalent, in form, content, and implication, to local justice.

Indeed, one initially plausible way of understanding the distinction between local and domestic justice is through the idea of coercion. Domestic injustices (those of the basic structure) might be suitable candidates for coercive intervention, whereas local injustices (those of informal associations or individual choices) might not (see Brettschneider 2007). This hypothesis leads to my second objection to Cohen: it is a mistake to think that an institution's coerciveness is decided before it is incorporated into the basic structure. Although I have cast doubt on what it means to say that the principles of justice apply to the basic structure, surely it must mean that an institution to which the principles apply thereby becomes a candidate for legal coercion. If some aspect of the family, for example, is part of the basic structure, then it follows that it might appropriately be the site of laws that ensure that it is structured in such a way as to instantiate the principles and ensure justice. Cohen has things the wrong way around: it does not make sense to think that we know that there must be laws before we know what they must be about. A liberal society enacts laws only if they are needed for justice. So an institution that is part of the basic structure might become a legitimate candidate for the application of coercion; it is not part of the basic structure because it was always, inevitably, the site of legal coercion.

Return to the example that Cohen gives: the sexist expectations that might attach to the roles of husband and wife. This injustice could be addressed in a coercive or noncoercive way. Coercive responses could include legally mandating certain sorts of equality (for example, entitling a housewife to half of her husband's salary, or requiring that fathers take compulsory paternity leave), changing the legal definition of marriage (removing any reference to obedience from any marriage recognized by the state, or even requiring that all such marriages include vows of equality), or measures that, while not directly coercive, are funded by taxation (state campaigns in favor of marital equality, or the compulsory teaching of sex equality in schools). Noncoercive responses would not involve the state; they might include private citizens, voluntary groups, or campaigning associations actively working to break down sexist stereotypes and offer support to couples trying to escape them. It would be possible for Rawls to maintain that the former, coercive options involved aspects of the family that were thereby part of the basic structure, and thus that the equality they brought about was part of domestic justice. The latter, noncoercive options would not be part of the basic structure and would belong to the domain of local justice.

The problem for Rawls, then, is not that there can be no meaningful distinction between justice secured by coercion and justice secured through other means. The problem is rather that whether justice is secured coercively does not depend on whether an institution is part of the basic structure. It is not because the family is or is not "part of the basic structure" that it becomes an appropriate or inappropriate site for justice "applying directly" to it. On this interpretation, whether or not something is part of

"The Family as a Basic Institution" 93

the basic structure is the result, not the cause, of its being an appropriate site of coercion.

This interpretation is meant to defend Rawls only from one specific part of Cohen's critique. As noted at the outset, I agree with Cohen that Rawls's version of the basic-structure distinction is doomed, and I agree with him also that noncoercive factors are crucial to justice.[14] But I do not agree that the latter explains the former. The basic-structure distinction is doomed because the family must be part of the basic structure, and because Rawls allows no way for the principles to apply "directly" to the family that he does not also allow for non-basic-structure institutions such as churches and universities.

Reconstructing Okin's Critique

Return to the formalization of Okin's argument that I gave above:

1. The family is part of the basic structure.
2. Rawls's theory entails that the principles of justice should apply internally to institutions that are part of the basic structure (but not to others).
3. Therefore, for Rawls, the principles of justice should apply internally to the family.
4. Rawls states that the principles of justice should not apply internally to the family.
5. Therefore, Rawls's position on the family is inconsistent.

We have now seen that premise 2 is false, and so conclusion 3 does not follow. Premise 4 is true, but without conclusion 3, conclusion 5 also fails. Nonetheless, I have not exonerated Rawls. We are in a position to substitute the following critique:

1*. The family is part of the basic structure.
2*. The principles of justice should apply directly to basic-structure institutions.
3*. Rawls states that the principles of justice apply differently to basic-structure vs. non-basic-structure institutions.
4*. Therefore, the principles should not apply directly to non-basic-structure institutions.[15]

5*. Rawls gives various examples of how the principles apply to the family, but these examples apply equally to non-basic-structure institutions such as churches, religions, and cultures.
6*. Therefore, Rawls is inconsistent.

Okin is right that Rawls's claims about the family are deeply confusing. The only way to make sense of Rawls's claims about the family is to put catastrophic pressure on the fundamental Rawlsian idea that the principles of justice apply to the basic structure of society only, or uniquely. Okin's work has successfully demonstrated that, if the idea of the basic structure is to have any weight, the family must be considered part of it. If Rawls wishes to maintain his claims about the limits of interference in the family, then he can do so only at the expense of the claim that the basic structure is uniquely the subject of justice.

Notes

1. My focus is on the complete picture of Rawlsian justice that we have at the end of his writing career, according to which Rawls advocates political liberalism, the political character of which depends in part on the idea that justice applies in a distinctive or unique way to the institutions of the basic structure.

2. For detailed discussion of the second problem, see Chambers 2008.

3. "I can see no good reason . . . to apply the difference principle to the property-holdings of a legislative body. . . . On the other hand, neither can I see any good reason . . . why the difference principle should not be applied within families" (Okin 2004, 1564).

4. For other claims that Okin's criticism of Rawls is problematic, see Lloyd 1995; Baehr 1996; Nussbaum 2000a; Wijze 2000; and the discussion in Abbey 2011.

5. This excerpt is very similar to material in IPRR.

6. Williams 1998 (229) also distinguishes between something being "within" and "comprising" the basic structure.

7. Discussion with Chris Brooke, Thom Brooks, Daniel Butt, Jon Quong, Ben Saunders, and Andrew Williams (to whom I am very grateful) reveals that there is significant controversy as to whether churches and universities are part of the basic structure. Additional textual support for my claim that Rawls excludes them can be found in JFR, 11–12. See also Freeman 2007, 101.

8. Abbey 2007 (17–18) refers to passages such as these as a "totalizing" view.

9. Of course, Rawls introduced the idea of the basic structure in TJ, before making the distinction between political and comprehensive liberalism, and so it is not clear whether he had that distinction in mind when writing about the basic structure. I find comprehensive liberalism more convincing than political liberalism. The fact that Rawlsian liberalism has difficulties maintaining its political credentials is, in my view, problematic for Rawls's own purposes, but not because comprehensive liberalism is implausible. Cohen argues that Rawls must maintain his focus on the basic structure if he is to retain his limited and distinctively liberal egalitarianism; without it, Cohen argues, Rawls becomes a "radical egalitarian socialist" (Cohen 2008, 129n27).

10. For example, see Lloyd 1995, 1327; Baehr 1996, 49–52; Nussbaum 2000a, 60; Wijze 2000, 274; and Ronzoni 2008. I am very grateful to Chris Brooke, Jon Quong, and Andrew Williams; each put forward versions of this answer in discussion and responded patiently to my queries.

11. Note, however, that the difference principle is not a good example of the claim that the family, specifically, could be unjust while the basic structure as a whole secures justice. For, as Okin has correctly argued, Rawls's insistence that the difference principle applies to heads of households rather than individuals is inconsistent with any defensible account of liberal values. In other words, the difference principle will have to reach inside the family to the individuals within it. This does not mean that the difference principle applies to the internal life of the family: the family does not have its own, self-contained difference principle. But it does mean that the difference principle cannot apply to the basic structure as a whole if it stops at the doors of the household. See Okin 1989a, 91–93.

12. See JFR, 10 and 164. I say "suggests" rather than "states" because Rawls cites business firms as organizations to which the principles of justice do not apply directly in a list that also includes the family. But this interpretation is supported by consideration of what *is* in the basic structure, for which, see the arguments above and also Freeman 2007, 101–2. I thus disagree with Stephen de Wijze (2000, 274), who places business firms within the basic structure.

13. Cohen could specify the first horn of the dilemma more precisely. Presumably, it is open to Rawls to say that all those features that have profound effects on people's life chances are part of the basic structure, and that justice applies to that structure. Cohen's objection would then be not that justice cannot be restricted to the basic structure but that the basic structure is more or less everything.

14. The purpose of Cohen's argument is to show that justice must require an egalitarian ethos from its citizens, such that incentive payments of a certain sort are not required. For further discussion, see Chambers and Parvin 2010.

15. The move from 3* to conclusion 4* could be questioned: it is possible that there is some other difference, not considered here, between the way the principles apply to the basic structure and the way they apply elsewhere. However, Rawls provides us only with the two ideas I have considered here: direct and internal application. Since the idea of internal application does not provide the necessary distinction, we are left with no viable alternative to the idea of direct application.

5

Rawls, Freedom, and Disability

A Feminist Rereading

Nancy J. Hirschmann

When political theorists and philosophers take up the question of disability, they generally do so with regard to questions of justice: *allocation* of resources to disabled people (whether directly through accommodation, subsidy, or health care, or indirectly, say, through scientific research); *distribution* of resources (which disabilities or illnesses should receive more dollars, which less; how do we decide from which other programs or areas to take resources in order to pay for that?); *entitlement* to resources (should public buses be fitted for wheelchair lifts if extremely few residents use wheelchairs?); and *adequacy* of resources (in granting disability payments through Social Security, how much is enough?). Questions of responsibility

---

Thanks to Ruth Abbey and Corey Brettschneider for comments and suggestions.

sometimes arise in terms of determining just distribution and entitlement (is a person disabled because of irresponsible action, like driving a motorcycle without a helmet?), but justice is the primary theoretical concept that one encounters in this disciplinary framework.

Much of this work revolves around, or was inspired by, the work of John Rawls. *TJ* captured the imagination of political theorists and philosophers like no other twentieth-century text except Michel Foucault's *Discipline and Punish*. Yet Rawls made little mention of disability in *TJ*; and in subsequent writings—in response to criticisms that he ignored the disabled, as well as women—the gestures he makes toward recognizing disabled people as equally entitled to justice are fairly inadequate. Indeed, he seems somewhat hostile in requiring that parties in the OP be "fully cooperating members of society over a complete life," which would thereby exclude people with "permanent physical disabilities . . . so severe as to prevent persons from being normal and fully co-operating members of society in the usual sense" (Rawls 1985, 233). More generally, he defines disability in terms of medical conditions that should be treated by health care (*PL*, 20, 184), thus misunderstanding the perspective of disability scholars and activists who view disabilities as physical differences that are discriminated against and should be accommodated by policy or law. True to the analytic philosophy tradition, Rawls justifies setting aside illness and disability because he believes that if we can establish principles along narrow idealist lines, we may be able to bring in more "complex" cases, which disability supposedly presents, at a later point (*PL*, 20; Daniels 1990, 278).

Is it possible that Rawls ignores disability because, like Habermas, he does not want to acknowledge the disadvantages to which his own disability—a stutter—subjected him? It is interesting that both Habermas's and Rawls's disabilities pertained to speech (for Habermas, a cleft palate led to unclear vocalization). According to some accounts, Rawls's stutter developed later in life, apparently in response to the deaths of two of his brothers, from diphtheria and pneumonia, and thus was highly atypical (Rogers 1999). Rawls never wrote about his stutter or was willing to consider himself "disabled." Nor did he write about his brothers in any of his brief remarks about health care.

A deeper explanation may be that disability poses such a radical challenge to Rawls's theory that he would have to rethink its fundamentals. As Eva Kittay (1999) suggests, justice requires that we include dependency needs not at a later, legislative stage that distributes resources to solve individual needs, but rather in the OP as a fundamental fact about humanity that must be included in the initial starting point. Martha Nussbaum

(2006) also identifies fundamental problems with Rawls's theory that exclude disability from the start. Deploying her "capabilities approach," Nussbaum maintains that justice requires a distribution of resources that aims at a threshold for developing the capability functioning of severely disabled individuals who are not able to meet Rawls's contractualist assumptions.

Other scholars believe that Rawls's theory of justice could, with some adjustments, accommodate disability. Cynthia Stark (2007) argues that despite Rawls's emphasis on reciprocity and cooperation, leading to the apparent exclusion of disabled individuals who, in his view, cannot "cooperate," his remarks about the social minimum nevertheless entail their inclusion. Jonathan Wolff (2009) proposes a strategy of "declustering disadvantage": specifically, in an affluent society, in which disabled people are likely to be among the worst off, Rawls's difference principle would be likely to mandate resources devoted to accommodation and incorporation of such individuals into the cooperative scheme (for instance, making expenditures that enable disabled persons to be employed). In less affluent societies, however, where disability is only one factor of many that determines the worst off, other mandates would take priority. Norman Daniels follows Rawls in taking a "biomedical model" approach to disability as the absence of health and as "deviation from the normal functional organization of a typical member of a species" (1990, 280). Daniels seeks to expand our understanding of the primary goods to include social support services and "preventive, curative, and rehabilitative medical support services" (280), and to widen Rawls's conception of "opportunities" beyond the economic and career-based to include health care (282, 292). Silvers and Francis suggest that contract theories such as Rawls's can be reformulated by replacing the "successful bargainer paradigm" with "principles of justice educed with an eye toward promoting a trust culture and a climate of trust" (2005, 43). This would eliminate the "outlier problem" by allowing a more inclusive conception of the parties to the OP—not only who can be a party to the contract, but how such parties can participate.

Most of these works do not really address Kittay's fundamental challenge that disability is excluded from the OP, and that the fundamental reality of human dependency is excluded from the two principles of justice. Even the most sympathetic of these works consider disability something that could be addressed at a later stage. Silvers and Francis, while claiming to address Rawls's fundamental assumptions, simply assert what Rawls denies: that contract is an appropriate mechanism for describing interactions among individuals who are not fully rational and reciprocal. They therefore do not confront the issue but rather contradict it.

More crucial to my own argument, none of these works, whether sympathetic to Rawls or critical of him, takes up freedom in the context of disability. All consider disability as exclusively a question of justice. Indeed, other philosophers and theorists who write on disability, even though they do not focus particularly on Rawls (see, among others, Anderson 1999; Arneson 2000; Feinberg 2000; Pogge 2000), similarly approach disability as a question of justice rather than freedom. Even Nussbaum's capabilities approach, which centrally addresses the relationship of capability functioning to women's freedom in *Women and Human Development* (Nussbaum 2000b; Hirschmann 2003), primarily considers capabilities within standard justice concerns of distribution, allocation, entitlement, and adequacy of resources, when it comes to disability, in *Frontiers of Justice*; at best, she is concerned with equality rather than freedom (Nussbaum 2006).

This is true of other feminist interpreters of Rawls as well; as is well known to readers of this volume, Rawls made very little reference to women, including them within families in *TJ* and positing that those occupying the OP were for the most part "heads of households" and thereby generally male. Though this stipulation seemed to have been dropped in favor of "individuals" pursuant to feminist criticism, such criticism of Rawls on this count has similarly focused on the injustice of his theory from a feminist perspective, particularly arising from Carol Gilligan's *In a Different Voice*. Gilligan's theory juxtaposed justice, understood as abstract principles, to an ethic of care based on concrete connection, relationships, and responsibility stemming from women's historical lived experience. In terms of applying Rawls's principles of justice to women's lives, Okin's critique in *JGF* has gained particular purchase, arguing that the vulnerability to which marriage exposes women requires that a theory of justice address the particular disadvantages to which this primary social structure subjects them. Her solution, in addition to laws against domestic violence and recognizing women's contributions to the household economy at the time of divorce, centered on the breadwinning spouse's wage being paid equally to him and his non-wage-earning wife. In other words, the problem of gender equality was a problem of justice that could be remedied by better distribution. There and in later work (e.g., 1994), Okin makes a broader argument that Rawls's failure to acknowledge injustice in the family has negative ramifications for women's public status. But she saw freedom as, at best, a secondary corollary of justice. Disability was not a topic that she took up at all (though her critique of Rawls's public-private dualism could be useful in analyzing the frequently ambiguous location of disabled individuals who may receive public goods like Social Security, or be institutionalized in public facilities, and yet are cut off from access to the broader public arena and denied political voice and power).

The few authors who have joined these two perspectives of feminism and disability, particularly Kittay and Silvers, have likewise critiqued Rawls's theory of justice rather than his theory of freedom. Particularly following Okin's and Gilligan's care-based critiques, they have argued that the notion of dependency is vital to the disability experience—indeed, to all human experience—and that justice has to account for it. Kittay (1999) argues that Rawls's criteria for justice that members of a polity be "fully cooperating," for those who are deeply dependent (which, by her definition, would actually exclude a fair number of disabilities, as she is concerned primarily with the severely disabled) as well as dependency workers, cannot be accommodated in the OP or, therefore, in the two principles of justice, because they violate the central conditions that define membership in the OP. Silvers (1995) is rather more critical of the care ethic for disability, and Elizabeth Anderson (1999) is more critical of egalitarian theories based on luck than of Rawls per se, but both engage Rawls and disability on questions of justice rather than on questions of freedom.

Yet freedom is arguably the, or at least a, central concept that underlies justice. Questions of injustice arise predominantly when people, disabled or not, cannot do what they wish; this leads to questions of remediation, which in turn lead to questions of whether remediation is justified and feasible, and in turn to questions of how resources should or can be allocated to produce the remediation. The question of freedom, on this reading, is the animating first question for justice. But freedom is in and of itself an important concept for all human beings, no less for the disabled than for the nondisabled.

I further maintain that Rawls is not simply a theorist of justice; he is also a theorist of freedom, and freedom is a central animating principle behind *his* theory of justice in particular. Though he claims that "no priority is assigned to liberty as such, as if the exercise of something called 'liberty' has a preeminent value and is the main if not the sole end of political and social justice" (PL, 291–92), this exaggerated disavowal does not undermine the fact that "the priority of liberty" does apply within the two principles. What makes something unjust, in his view, entails the inhibition of people's liberty in various ways, and Rawls's theory of justice intends to compensate for, if not overcome, the unequal distribution of freedom that results from such inequalities of ability. His fundamental premise that we do not "deserve" the natural talents or disabilities with which we are born any more than we "deserve" the class, culture, and status in which our families are situated provides a launching point for addressing the social and biological disadvantages in which certain physical states in particular contexts place us. The most serious limits to his theory in terms of disability stem from its underlying theory of freedom, which encodes a notion of "embodied individualism" that has characterized liberal theory since at least Thomas Hobbes (Hirschmann 2013). Though the "veil of ignorance" and the OP posit essentially bodiless rational minds, Rawls in fact draws implicitly on certain substantive assumptions about what kinds of bodies not only are but can ever be free within the terms of liberal discourse, and hence appropriate subjects of justice. That is, Rawls's theory of freedom is not only a masculinist theory of freedom, but it is also an ableist one.

Rawls's Masculinist Theory of Freedom

Rawls articulates his conception of freedom in the first principle of justice: "Each person is to have an equal right to a fully adequate scheme of equal basic liberties which is compatible with a similar scheme of liberties for all" (PL, 291). In TJ he invokes Gerald MacCallum in claiming that all instances of freedom "can always be explained by a reference to three items: the agents who are free, the restrictions or limitations which they are free from, and what it is that they are free to do or not to do" (202), thereby claiming to sidestep Isaiah Berlin's division between negative and positive liberty. But in fact he offers a simple articulation of the negative liberty principle in TJ, particularly in his account of the equal basic liberties, which he modifies in PL to "freedom of thought and liberty of conscience; the political liberties and freedom of association, as well as the freedoms specified by the liberty and integrity of the person; and finally, the rights and liberties covered by the rule of law" (291). And of course a centrally important liberty of the person "is the right to hold and to have the exclusive use of personal property" (298).

Certainly, these freedoms are valuable to many disabled individuals; my point is simply that they are all central ideals of negative liberty, and Rawls offers a classically liberal articulation of negative freedom. He offers no account of any of the central ideals of positive liberty articulated by theorists from Rousseau to Charles Taylor. Consider the place of personal development, for instance; though pursing "life plans" is important to Rawls's theory, such life plans are so only as a *product* of freedom, not as part of its meaning. That is, freedom from restraint is important so that I can pursue whatever life plan I wish. But the need to develop my person is not part of the meaning of freedom; it is only something that I should be free to do if I wish. Community, similarly, is merely a cooperative venture among free individuals for Rawls; it does not produce my capabilities but rather is a function of the basic capabilities I already have. In Rawls's view, relationship or community is something that free individuals choose freely to form, not something that produces my preferences and even my ability to choose. Similarly, Rawls offers no account of internal barriers to freedom, the intense immediate psychological dispositions and desires that pull on me to violate the longer-term desires that are more important and valuable to me.

Indeed, Rawls explicitly *excludes* all of these things from freedom, saying that they do *not* pertain to freedom per se but to "the worth of liberty" (*TJ*, 204). Anything that the positive libertarian might include in her formulation—such as equality, or wealth, or strength—falls, for Rawls, outside the definition of liberty and is included in something else: the worth of liberty, equality, or justice.

As feminists have argued, this conception of freedom is fundamentally masculinist. First of all, the "individual" at its heart is one who tacitly depends on others, particularly women, to do the work of caring and of building relationships, and the tasks of affective production, even as it denies this dependence. As feminists since Simone de Beauvoir have argued, such labor is what makes it possible for others, generally men, to engage in a wide variety of activities—what Beauvoir called "the world of transcendence"—associated with modern conceptions of freedom, such as creativity, production, and rational thought. The fact that affective production makes such activities possible in the first place is simply suppressed. Moreover, theorists like Gilligan have argued that it is women's attention to the realm of care that allows men to develop and believe in a conception of justice that operates on a foundation of rules, principles, rationality, and, of course, the logic of individual free choice, particularly contractualism. Rawls's, of course, is the ultimate contractualist justice theory, and indeed, Gilligan's mentor, Lawrence Kohlberg—whose work Gilligan's theory critiques—explicitly acknowledges Rawls and Rawlsian justice in the "stage theory of morality" that he develops and that Gilligan rejects (Kohlberg 1981; Hirschmann 1992, chap. 3).

Finally, Rawls never addresses the most important dimension of positive liberty, namely, the social construction of desire. How is it that we want the things we want, that we have the desires we have? The idea of social construction holds that human beings and their world are in no sense given or natural but are the product of historical configurations of relationships. Our desires, preferences, beliefs, values—indeed, the way in which we see the world and define reality—are all shaped by the particular constellation of social relationships that constitute our individual and collective identities. Understanding them requires that we place them in their historical, social, and political contexts, which make meaning possible.

Rawls ignores social construction in his conception of freedom. In *PL* he does identify "political constructivism" as an important element of his theory; but this entails only a recognition that the procedure by which we choose the two principles stems from the artifice he creates, the OP. Hence he says, "what is it that is constructed? Answer: the content of a political conception of justice. In justice as fairness this content is the principles of justice selected by the parties in the OP as they try to advance the interests of those they represent" (*PL*, 103). This statement bears no relation to the idea of social construction I articulated above. Moreover, Rawls declares that the OP itself is not a construction; "it is simply laid out" (103). But social construction theory maintains that what made it possible for Rawls to think of this idea was his particular location within the matrices of gender, race, and class that defined his life. The widespread character of these matrices may explain the great appeal of his theory to a certain subset of academics who are similarly located; and indeed, the fact that feminist and critical race theorists are almost uniformly critical of Rawls could be the result of their quite different location within those matrices. Moreover, according to Rawls, the subject of this "constructivism" is a principle of justice, which is selected by "citizens," an abstraction that considers irrelevant the particularities of the human beings occupying (and excluded from) that role.

Accordingly, Rawls notes that "when citizens convert from one religion to another . . . they do not cease to be, for questions of political justice, the same persons they were before" (*PL*, 30), a claim that social constructivism would clearly dispute. Religion, like gender, sexuality, and race, can be viewed as constitutive of our "personhood," and these categories are all constructed by the particularities of history, context, and culture. Women in radical fundamentalist religions, for instance, may have a substantially different identity—personally, socially, and politically—than they do in more moderate religions (Ahmed 1993; Hirschmann 2003). And this difference in identity is tied up with how women see themselves as well as how others see them.

Far more significant is Rawls's account of the "free citizen" in *PL* and *JFR*. The first condition, that citizens "conceive of themselves as having the moral power to have a conception of the good," could bolster disability claims as long as such claims are "reasonable"—an assertion that disability activists have learned to treat with deep suspicion. More problematic is the argument that they "take responsibility for their ends," an idea with at least ambiguous meaning for severely disabled people, and even for moderately disabled people who need assistance from others in communicating with nondisabled persons. But perhaps most problematic is Rawls's claim that the parties in the OP are "self-authenticating sources of valid claims" (*PL*, 32; *JFR*, 23); that is, individuals' claims are valid if they say so, nobody else can judge the validity of my desires. The nature of desire is that it is intrinsic to me: I want what I want, for whatever reason I want it. The dictates of justice may or may not entitle me to social assistance in *fulfilling* my desires, and may thereby make it more or less difficult to *pursue* my desires. But my desires or claims themselves need not adhere to any external source of validation.

My aim here, therefore, is not only to raise questions about whether disabled people can be "free citizens" in Rawls's formulation but also to highlight the fact that Rawls takes no account of social construction. He assumes a unitary and uniform perspective of desire. The difference principle, for instance, which claims that I will be content to accept a worse-off position in an unequal relationship if the inequality makes me better off than I would otherwise be, assumes that desire is uniform and static. The OP is structured to eliminate all knowledge of my particular social history, my relationships, my cultural practices—all of the things that make humans what they are, that socially construct us. Desires therefore must always fit within the framework of rational self-interest in Rawls's view. Furthermore, the conception of "rationality" that Rawls deploys is decidedly a liberal one; the primary goods that he articulates are defined from a liberal perspective and rule out other sorts of fundamental goods, such as community and culture—things that, in his view, enter only at later stages under the fundamental "freedom of association," as if cultures and communities are things that people pick and choose and put together *within* the framework of their lives rather than preexisting and deeply shaping that framework. In this sense, Rawls adopts a classically liberal abstract individualism as the foundation of his theory of freedom: as in social contract theory, the isolated individual is the center of the universe and the basic foundation; that individual decides which relationships and connections to forge and thereby creates his or her relationships and communities, which are thereby contractually produced on the basis of self-interest.

As I suggested above, Kittay argues that Rawls's theory "is shaped without attention to the role of dependency in our lives," and she says this also about Rawls's "norm of freedom" (1999, 96). But I would push that claim further; for the standard account of freedom, like his, does not take into account the role of *relationships of all kinds*. Rawls, like most other liberal theorists, sees relationships as things that individuals can create, choose, and consent to, once the conditions for individual freedom have been set. Indeed, this premise is a central dimension of the social contract theory that so deeply informs Rawls's theory: it is only after the assertion of individual freedom that the individual's ability to "construct" society via contract is possible. For Rawls, justice is the core of that contract, just as property is the core of the contract for Locke.

Social construction is important to freedom on several levels, the most obvious being that the options and opportunities that are made available to people and from which they can choose are often a function of social policy and social choice. But what positive liberty theory further shows is that a lack of options affects desire itself; who we are, and the possibilities for expressing desire, even for having desire, are things that are shaped and produced by the choices available to us concerning the material conditions of our lives. Feminists have argued that understanding the way in which desire is produced and structured along gendered lines is essential to understanding freedom. The ways in which gender is structured as a social concept ensures that certain socially contingent and socially produced preferences and desires are normalized and naturalized for men and women in ways they are completely unaware of. Thus we take as natural and given things that are in fact socially produced, culturally specific, and historically contingent. For example, Kittay notes that dependency workers—particularly mothers—end up "choosing" to care for their children because of the lack of other viable options, and that "because care of dependents is nonoptional in any society, some societal measures are inescapably taken to meet the inevitable need for care" (1999, 99). She implies that the curtailing of women's options has historically been a key measure so taken, ranging from the naturalization of women's caretaking roles to the explicit prohibition of their participation in the public sphere; women are left to do that work because others will not. "The consequence," Kittay concludes, "is that many claims are presumed to be self-authenticated when they are really heteronomous" (99). In other words, women's desires are constructed by and through the social limitations to which they are subjected.

Insofar as Rawls's theory collaborates with this construction, as feminists have shown, it collaborates with this denial of freedom. Rawls's central emphasis on a particular mode of rationality as key to justice, his reliance on rules rather than relationships and connection, the disembodiment of persons in the OP—all of these aspects of Rawls's theory of justice, which feminists have critiqued, similarly reveal his concept of freedom as masculinist.

The Ableist View of Freedom

I further contend that Rawls's theory of justice is also "ableist," or biased against persons with disabilities. Social construction takes on particular significance for disabled persons. Take, for example, a person who uses a wheelchair and wishes to attend my class. The building in which the classroom is located has no ground-level entrance, no ramp, and no elevator. This student is therefore unable to attend my class. Does this inability constitute a lack of freedom? And, if so, is it an injustice?

What is called the "medical model" of disability leads to the standard philosophical answer: no. The "medical model" views the student's problem in terms of her bodily incapacity; her body's inability to climb stairs prevents her from entering the building. Her incapacity is considered an individual medical problem; she might suffer from this condition along with other individuals, but each of them has an individualized medical condition that is unrelated to the conditions that other similar sufferers have. Moreover, *sufferer* is a key term, for all such conditions are viewed as tragedies of bad luck, conditions from which the affected individuals must want to escape, even if that is impossible, and which negatively affect their quality of life.

This is the dominant model presumed by many philosophers who write on disability, and even more by those who write on freedom. In particular, it is an accepted view of contemporary theories of freedom that freedom presupposes ability. Richard Flathman offers the standard example: we are not able "to jump, unaided, twenty-five feet straight up from the surface of the earth, to develop gills instead of or in addition to lungs. . . . Few if any of us decide to stand upright, to walk by moving first one foot and then the other, or to see figures three-dimensionally." What humans are able to do determines the context for freedom, for even thinking about freedom: "Conduct that is mutually meaningful—and hence *possibly* freedom-evaluable—is conduct within these general facts as they are accepted by the 'reasonable' participants in a form of life. These facts, and the limits they 'impose' on our conduct, are for the most part beyond our powers to change" (Flathman 1987, 139). Therefore, they do not really "impose limits" at all but merely define the parameters of possible action. If I am unfree, then someone—an agent—must be preventing me from doing what I want (or forcing me to do what I do not want), and this agent must be acting purposively and intentionally. If purposive and intentional restraint is missing, then we cannot say that I am "unfree" (see also Benn 1988).

Though particularly true of theorists of negative liberty, these assumptions are also found in some who advocate positive liberty. For instance, Kristjan Kristjansson offers the example of someone "confined to a wheelchair" because of a broken leg, saying that in such a case the person "can be considered free (though unable) to run—unless there is, for instance, a law against running, in which case this person would be unfree to run in addition to being unable to do so" (1992, 297–98). That is, no one is preventing such a person from running (unless it is against the law; or unless someone has purposely broken his leg so as to prevent his running). Rather, he simply lacks the ability. Disabled persons, on this view, are not the appropriate subjects of liberty; liberty is not relevant to them because others are not preventing them from doing things they wish to do; they are simply unable to do them, not unfree.

When Kristjansson hypothesizes a broken leg—a condition from which he will recover within weeks—rather than a permanent disabling condition, he invites the reader's sympathies to agree with his argument: the broken leg is, we think, an inconvenience that many people experience at some point. But our perspective may shift if we think of my earlier example. The student using the wheelchair is, first of all, not "confined" to it; the wheelchair is the means of her mobility, the way she moves (and notice that although Kristjansson uses the phrase "*confined* to a wheelchair," his subject is still "free," an apparent contradiction). Second, she is not concerned with running but rather with entering a building in which her first class as a newly registered university student is being held. So what she is prevented from doing is not simply a particular act but rather is tied up with an entire life plan for her future. Moreover, it is not a life plan that is eccentric or difficult to fulfill but one that vast numbers of other people pursue every day. On the standard formulation, there is no assignable agent preventing her from entering the building, so she would be unable but not unfree to attend the class; and her inability would seem to be "internal" to her, constituting the parameters of her ability. Therefore, the stairs do not constitute a barrier, for stairs were not created—by their original inventor, or by the person who designed this particular building—with the idea of keeping out people who use wheelchairs. Indeed, the reason why wheelchair access is so uncommon is that disabled people rarely even entered the consciousness of designers, architects, and builders before the Americans with Disabilities Act (ADA) was passed—and not even then, in many cases.

Two underlying assumptions are at work here. The first is that "the world as we know it" is conceived as in a sense "natural," not as a product of agency and choice; humans cannot be blamed for building stairs thousands of years ago, before wheelchairs were even invented; stairs were a part of the "natural evolution" of the human practice of building. Although cost is the primary justification for exempting many inaccessible older buildings from retrofitting, the tacit moral argument underlying that justification is that nobody intended to harm disabled people via such architectural design. (Consider, by contrast, that when it seemed that tobacco companies did intend harm to their customers by hiding the results of their research, they were expected to pay very large sums of money regardless of the cost.)

The second assumption is that barriers to freedom must by definition come from outside the self. This is a classic aspect of negative liberty, where freedom is defined as an absence of external constraints or barriers to doing what I want. This is where the medical model of disability is particularly problematic from the perspective of freedom, for within its framework the "barrier"—what prevents the person from acting—is seen as coming from inside the self. By contrast, the "social model" of disability maintains that while impairments such as blindness or paraplegia may be biologically based, the notion that these impairments constitute "disabilities" is produced by social arrangements that favor certain bodies and penalize others. I may be unable to walk, but the lack of a wheelchair, or the existence of buildings with stairs rather than ramps, "disables" me from moving or from entering a building. As Silvers says, "it is the way society is organized rather than personal deficits which disadvantages this minority" (1996, 210).

Disability is thus a "social construction" in the most obviously political sense of that term. What makes a bodily condition a "disability," on this view, *does* pertain to external factors: it is not the spinal cord damage affecting my legs that is relevant but the stairs, which are an external barrier to my freedom. Similarly, the height of sinks, counters, elevator buttons, and so forth is a result of human choices that effectively exclude people with a variety of physical differences, such as dwarfism, paraplegia, and so forth. The presence or absence of braille or telecommunication devices for the deaf reflects a choice that determines the inclusion or exclusion of people with vision or hearing impairments. Such prior choices determine an opportunity set that limits certain individuals.

Although many philosophers do not see disability in this sense, seeing it rather as an intrinsic inadequacy of particular bodies, this social view of disability as produced by the built environment and social context is at the heart of the ADA: "Individuals with disabilities . . . have been faced with restrictions and limitations, subjected to a history of purposeful unequal treatment, and relegated to a position of political powerlessness in our society, based on characteristics that are beyond the control of such individuals and resulting from stereotypic assumptions not truly indicative of the individual ability of such individuals to participate in, and contribute to, society" (Americans with Disabilities Act of 1990, 42 U.S.C. §12112). Disabled persons have fewer choices than the nondisabled, all things considered, because they are actively excluded from buildings, schools, and jobs; they generally have lower incomes as a result and therefore have diminished options generally associated with wealth. In short, they have considerably less of what Rawls calls the primary goods—but this lesser amount is due to social choice, not to their own deficiencies per se or to "bad luck." The barriers to their freedom are thus, in this view, external, not internal.

Many disability theorists and activists go further than this. In their view, "impairments" are simply another form of "difference." Advocates for the "Deaf," for instance—purposely deploying the capital "D" (contrary to many publishers' rules!) to communicate that deafness constitutes a culture with its own language and practices—maintain that deafness is not an impairment or a disability in any sense of the words. They claim that sign language is simply a different form of communication and that deafness itself is a cultural difference. What turns deafness into a disability is the prejudices of the hearing world, which has over time treated deaf people in a variety of negative ways, ranging from institutionalization for supposed mental deficiencies, to prohibitions on sign language and forcing deaf people to read lips and speak, to the new efforts by hearing parents to give cochlear implants to their deaf children. Similarly, Temple Grandin has argued that autism is not a cognitive deficiency but a cognitive difference, enabling her to see the physical and social world in different, but no less valuable, ways from the "cognitively normal." In Grandin's case, "seeing in pictures" enabled her to develop a method of cattle slaughter much more humane than conventional methods and much less cruel and stressful to the cattle (Grandin 1995). (I take no position here on the question whether killing animals can ever be an ethical act.) Grandin's cognitive difference is thus not an intrinsic disability; it was made into a disability by those who failed to understand and appreciate her contributions.

On this reading, then, the social model of disability can seem to conform quite well to a negative-liberty view of freedom; by seeing the ways in which conditions that we have considered "natural" or "inevitable" are in fact produced by humans and interfere with the liberty of disabled individuals, we could use negative liberty to expand what "counts" as an external barrier to freedom. Feminism, too, can accommodate itself to a negative-liberty view of freedom, as I have previously argued: not only are the standard views of barriers such as sexual discrimination, sexual harassment, and sexual assault, as well as legal restrictions on abortion, education, and employment, all important to feminist arguments about women's freedom, but the larger argument that patriarchy itself, and all of its norms, customs, rules, and practices, is an external limitation on women's freedom is similarly vital to feminist understandings of freedom (Hirschmann 2003).

But while such an expansion of negative liberty is necessary for both feminist and disability perspectives, it is not sufficient. For the social construction of disability, like that of gender, goes further than the material level of social choices about the built environment. It pertains to the normative content of our thoughts about people with bodily differences. The notion that blindness or deafness or spinal cord damage is a supreme loss to the individual, a disaster that the individual would want to overcome at all costs, constructs "the disabled" into a tragic figure, dependent, sick, weak, and unable to make any but the barest contributions to the collective social welfare. The terminology used to describe such people—handicapped, crippled, disabled, impaired—invokes loss, deprivation, tragedy, and subhumanity. Even Daniels, who argues that justice requires a system of universal health care, treats disability not as a potentially valuable difference but as a medical disorder that is a function of "bad luck" (Daniels 1985). And of course "resource constraints" feature significantly in the question of how much health care people are entitled to—again suggesting a medical rather than a social model of disability. The ways in which disabled people are thought about—the assumption that they must be very unhappy, and that one would, oneself, prefer to die than to be blind, or paralyzed, or to have MS—constructs policy decisions about, and daily treatment of, the disabled. Such conceptualization affects desire in different and deeper ways, profoundly shaping people's sense of self and subjectivity.

For instance, people who use wheelchairs do not generally ask to be "cured," and walking is not the necessary goal; movement is. And what movement requires is not "health care" (though that is important to everyone) but rather a good wheelchair and an accessible environment. Though much is made of the expense of putting in curb cuts, elevators, and ramps, studies show that many ambulatory people use these as well, as do people with disorders that are less severe. David Mitchell relates a story of how, when exiting the train en route from his home to his office at Temple University in Philadelphia, most passengers would go one way, to take the stairs, and he would navigate with his wheelchair on a circuitous route off to the side. When a ramp was installed in the station, what struck Mitchell was not the convenience or the time he saved but the fact that *most of the other passengers* walked up the ramp rather than using the equally nearby stairs. Disabled individuals ask that the "difference" of their conditions not be exaggerated and exoticized, as if accommodating them required turning the world upside down. Such exaggeration serves as a justification for the failure to remove barriers to their freedom.

When philosophers see disability in terms of "the medical model," and focus exclusively on questions of distributive justice, the kinds of expenditures required to deal with disability seem excessive to (mostly nondisabled) philosophers, just as philosophical thinking that has been done predominantly by men misunderstands, misconstrues, and misrepresents many aspects of women's lives. By contrast, looking at disability from the perspective of freedom makes "the social model" of disability much more logical. "Accommodation," the adaptation of the built environment to provide universal access, is all that many disabled individuals need in order to pursue basic activities such as working, living in their homes, getting dressed, cooking in their kitchens, and going on vacation.

Indeed, expenditures on some adaptive technologies—wheelchairs, for instance—would be paid back by a market economy in which such items would be purchased; think of the proliferation of "scooters" for senior citizens. Most large grocery stores have a few of these for their elderly and other mobility-impaired customers because it makes good economic sense for them. Graham Pullin (2009) has argued that wheelchairs and other adaptation devices could be designed much more effectively, efficiently, and beautifully, making them more desirable; the actress and model Aimee Mullins, for instance, owns several different prosthetic legs, including a pair made of glass and one of intricately carved wood. What is important to her is not "a desire for perfection [but] a desire for options" (Mullins 2009). Hearing aids, instead of becoming smaller and smaller, verging on the "invisible," and in the process working less effectively, could become larger again if they were designed in a visually attractive way, thereby lessening the stigma and encouraging more people with borderline hearing impairments to wear them (Pullin 2009). This would enhance freedom by, for instance, improving hearing-impaired people's ability to participate in conversations in crowded rooms.

Ableism in Rawls

Rawls might not say that the disabled have less freedom than the nondisabled, but rather that the freedoms they have are worth less than they would be if disabled persons were able-bodied. But such a view accepts as given the limitations of the built environment, the biases against certain cognitive orientations (such biases, indeed, are seen as rationality, not as biases at all), and so forth. Moreover, the medical model of disability imports certain assumptions into its evaluation of what justice requires that distorts the freedom claims of disabled individuals. When Rawls limits his understanding of disability to a medical view of bodily abnormalities caused by "accidents and illnesses," with the appropriate response being to "restore people by health care so that once again they are fully cooperating members of society" (*PL*, 20, 184), he profoundly misunderstands disability and, in the process, undercuts freedom for disabled people.

Rawls's ableist views lead to his failure to conceptualize adequately how a theory of justice could address questions of disability, and I maintain that this is because he fails to appreciate the role of freedom in such questions. Even adding disability to the conditions that we don't know about in the OP, along with religious beliefs or level of income, is inadequate, because studies regularly show that nondisabled people are unable to imagine the reality of disabled experience and hence to imagine what they would need if, once the veil of ignorance was lifted, they were disabled. For instance, even though disabled people report levels of well-being and happiness roughly equivalent to those of nondisabled people, the latter routinely report that, were they disabled, they would anticipate much lower levels of welfare and happiness (Brickman, Coates, and Janoff-Bulman 1978; Ubel, Loewenstein, and Jepson 2003). Moreover, able-bodied people simply *disbelieve* empirical evidence that people with disabilities are happy, much less that they would not prefer to be able-bodied (Weinberg 1988). As Silvers notes acerbically, "If . . . 'normal' individuals cannot accurately estimate what they would be like if in a 'damaged' individual's place, then truths about how one would want to be treated if one were disabled are opaque to 'normal' individuals and cannot motivate them morally" (1995, 36).

It is possible that if such information were presented to the people in the OP, they might think differently about disability, its inclusion in the two principles of justice, application of the difference principle to disability, and so forth. And of course Rawls's conception of "public reason" demands that all offer reasons to one another that respect their mutual status as equal and free citizens. But even this depends on "a principle of reciprocity, giving the same types of reasons to others that they can expect in return," which simply replicates the same exclusionary modes that feminists have critiqued in the case of gender (Brettschneider 2007). As Silvers observes—about "modern moral thought" more generally, but I think she must have Rawls's OP particularly in mind—"Far from extending protection so that it cloaks even individuals with disabilities," it "tends to magnify the influence of the medical model, with the result, inexorably, of excluding these individuals from normal moral recognition" (1996, 221). The problem goes to the root of who the "people" in the OP are in the first place. As various scholars have pointed out, they are likely to be white, most definitely male, and, at least in *TJ*, heads of households. They may be poor, given the amount of attention Rawls devotes to the problem of economic inequality and the need to reconcile equality with overall productivity levels, not to mention with freedom. But they are most certainly able-bodied and cognitively "normal."

Feminists have long shown us the importance of bodies to political theory and philosophy, which for centuries have operated on an arrogant presumption that the mind can be disembodied, that the body is particular, while the mind can appeal to the abstract and general, that the body is inessential, while the mind is the core of our humanity. Such beliefs, however, are possible only for certain kinds of bodies that can be taken for granted: not bodies that can get pregnant, not bodies that are tortured and killed because of their skin color, not bodies that are kept out of buildings because of architecture. For all the efforts to move away from the body that has characterized political philosophy, culminating in Rawls's OP, the embodied individual at the heart of liberalism is an able-bodied one.

As Peter Handley notes, Rawls "appears to question the extent to which disabled people might possess the requisite capacities for modern democratic citizenship on the basis of their perceived limitations" (2003, 111). Once again, consider Grandin, a woman who faced both gender and disability discrimination throughout her life. Cognitive disabilities such as autism are viewed as "deficits" within the Rawlsian frame, and individuals with such "deficits" are not considered "normal and fully co-operating members of society in the usual sense." But I hope I have shown that this "sense" is only "usual" from a particular masculinist, able-bodied perspective. For Grandin's work shows that it is precisely the *unusual* ways in which such individuals think that determine the ways in which they live their lives and the contributions they might make to the social whole. Grandin spoke, in essence, a different language, which required that she spend much more time, effort, and energy to engage with "normal" society, and to have her ideas understood. She wanted to make a valuable contribution to the "cooperative scheme," and indeed her "abnormality" is what made that contribution possible; yet in order to make it she first had to overcome barriers created not by her condition but by others' perception and (mis)understanding of it, and by their prejudice and fear of what they did not understand. Rawls's conception of justice would grant that Grandin is a "cooperating member," but only *after* she had suffered through the extra trials and demands just to be permitted to try, only *after* her efforts had failed because of others' prejudicial behavior. By the time she could enter Rawls's scheme, the injustices would have already been set, much as Okin (1994) argues about women. As Okin points out, the injustice of "background" conditions that make social cooperation possible, and hence "justice" itself, is something that Rawls's theory fails to confront, whether the background is the privilege of white masculinity or of able-bodiedness. Those conditions, and Rawls's ignorance of them, have their roots in a masculinist and ableist theory of freedom.

6

Rawls on International Justice

Eileen Hunt Botting

In *Frontiers of Justice: Disability, Nationality, Species Membership* (2006), Martha Nussbaum advances a feminist critique of Rawls's theory of international justice. While she identifies herself as a student of Rawls and dedicates the book to his memory, Nussbaum seeks to show the insufficiency of his theory of international justice, particularly as found in his *LP*, when it comes to defending the rights of women in developing countries. She contends that this insufficiency is mainly due to his use of social contract theory, and its attendant concepts of basic human rights, nationality, and the public-private distinction. In contrast, she offers an alternative human rights approach to global justice: the capabilities approach. The capabilities approach seeks to establish and defend a just and robust standard of human development across nation-states. While the capabilities approach gives Nussbaum an edge over Rawls in making strong, universalistic critiques of unjust patriarchal, religious, and familial practices toward women in developing nations, it runs the risk of *seeming* imperial and *appearing* as though it seeks to impose "Western" and "feminist" values on such nations. For this reason, the universalistic language of Nussbaum's capabilities approach is more valuable for defending women's human rights in situations of minimum cultural or religious conflict. Because of his stricter commitment to tolerating a reasonable pluralism of values across nations, Rawls's basic human rights approach is more useful for defending women's human rights in situations of cultural or religious conflict. Moreover, Rawls provides an argument for the just emergency use of military force to defend women's human rights, a hard case of which Nussbaum's theory steers clear. Through the comparison of their theories of justice, both Rawls's and Nussbaum's human rights approaches emerge in different ways as useful resources for advancing feminist values in international and transnational relations.

International Justice in Rawls

Rawls begins LP with a statement of the book's purpose, namely, to defend "a particular political conception of right and justice that applies to the principles and norms of international law and practice" (3). He proceeds to define the scope of his theory of international relations, stating that this "Law of Peoples" applies to the members of the "Society of Peoples," who follow its principles in relation to one another and other types of nations (3). This "Society of Peoples" includes peoples with liberal constitutional democracies as well as nonliberal "decent" peoples. He specifies one type of decent peoples, "decent hierarchal peoples," which he defines as those nonliberal, nonimperial nations that at least meet a minimum threshold in respecting basic human rights, including women's rights, and in consulting their people in political decisions made within a hierarchal political structure (4).

Rawls outlines a two-step philosophical process for the extension of liberal political principles to the law of peoples adopted by liberal and decent peoples. Using social contract theory as his guide, he envisions a hypothetical two-stage social contract. The first stage reaches agreement on principles of justice to govern the basic structure (the political and economic life) of a liberal people. As described in *TJ* and *PL*, a liberal people comes to accept two principles of justice—known as "justice as fairness"— that are generated from a thought experiment.

In this first-stage thought experiment, or OP, a set of free, equal, rational, self-interested, mutually disinterested parties deliberate about justice behind a "veil of ignorance" that screens out knowledge of their social positions, natural advantages, comprehensive doctrines (complex metaphysical and religious worldviews), and the distribution of wealth in their society (TJ, 12). This "veil" allows for fair and risk-averse deliberation about principles of justice for the basic structure of a liberal society, because it prevents the parties from making decisions that are partial to their own cases. Behind the veil, the parties deliberate about which principles of justice would serve their liberal society's commitment to the values of freedom and equality. They decide on two: the first principle protects everyone's equal right to civil and political liberties that are compatible with similar liberties for others; the second principle ensures (a) a fair equality of opportunity in access to social and political positions, and (b) a fair gap in income and wealth between rich and poor, such that the poor benefit from any economic advance by the rich. The parties order the principles serially, so that equal civil and political rights are protected above all else, and fair equality of opportunity takes precedence over economic fairness. The two principles of justice aim to establish a fair distribution of "primary goods" (rights, opportunities, income and wealth, self-respect) such that all members of a liberal society, regardless of natural advantages or socioeconomic status, may choose and pursue lives of self-respect governed by their own conceptions of the good life.

Liberal peoples come to accept and implement these two principles of "domestic" justice because they resonate with their "considered judgments" (or established values) about justice (TJ, 19). In particular, the two principles coincide with the liberal sentiment that justice means fairness in both procedure and outcome. In addition, individuals in a liberal society come to see this conception of "justice as fairness" as at least partly overlapping with the moral values embedded in their own, deeper metaphysical conceptions of the good life, if they hold reasonable comprehensive conceptions of the good life. This builds an "overlapping consensus" between "justice as fairness" and the variety of "comprehensive doctrines" (or complex religious or metaphysical worldviews) that individuals hold within their liberal society (Rawls 1985, 245–51). This "overlapping consensus" produces a stable agreement among liberal citizens about the value of upholding "justice as fairness" as governing the basic structure of their society (246–47).

Rawls describes this kind of liberalism as "political liberalism," in contrast to a kind of liberalism with deep metaphysical commitments, say, to a theory of the person, human nature, or a thick conception of the good life (*PL*, xviii–xxi). A political liberalism tolerates a diversity of reasonable comprehensive doctrines among its citizens while neither relying on nor endorsing any metaphysical view as the basis of its own theory of justice. At the same time, politically liberal societies understand these comprehensive doctrines as important for giving meaning and self-respect to the lives of individual citizens, and for supporting and stabilizing popular agreement on "justice as fairness" as their framework for governing the basic structure of society. For these reasons, a political liberalism remains neutral toward religion, allowing freedom of conscience and disallowing state establishment of religion, but supporting religious pluralism and toleration in civil society (*LP*, 55).

The first part of *LP* concerns the second-stage hypothetical social contract that liberal peoples, governed by "justice as fairness" at the domestic level, would use to reach agreement on principles of international justice among them. Rawls proposes that in this second-stage thought experiment, the parties in the OP are the "rational representatives of liberal peoples" (32). They are modeled as free and equal parties to the deliberation over the "Law of Peoples"—or the principles that govern "the basic structure of the relations among peoples" (33). The representatives in the second-stage OP are modeled as "free and equal" because "we view peoples as conceiving of themselves" as free and equal with respect to each other in the "Society of Peoples" (34). This is a far weaker and more pragmatic claim than insisting that the members of the society of peoples are actually equal in economic or political might, really equal in liberty, or truly free.

In order to ensure fairness in the procedure of the thought experiment, the veil of ignorance screens out the parties' knowledge of "the size of the territory, or the population, or the relative strength of the people whose fundamental interests they represent," as well as their "natural resources" and "level of economic development" (32–33). They do know, however, that "reasonably favorable conditions that make constitutional democracy possible" obtain in their countries (33). Because the parties are hypothetical representatives of politically liberal peoples, the veil also screens out knowledge of any comprehensive doctrines that individuals may hold in liberal societies (34). With the veil securing conditions of fairness for decision making, the parties select principles of international justice based on the "right reasons": their "fundamental interests" as liberal peoples as defined by their domestic-level principles of justice (33).

The deliberative outcome of this second-stage hypothetical social contract is the eight principles of the law of peoples. First, peoples are "free and independent" and are to be respected as such by other peoples; second, peoples should "observe treaties and undertakings"; third, peoples are equal as well as "equal parties to the agreements that bind them"; fourth, peoples should observe "a duty of non-intervention" in other peoples' domestic structures; fifth, peoples have a "right to self-defense" but no right to war beyond self-defense; sixth, peoples must "honor human rights"; seventh, peoples must honor rules of engagement in war; and eighth—and most controversially, according to Rawls—peoples have a "duty to assist other peoples living in unfavorable conditions that prevent them from having a just or decent social regime" (37). Rawls does not serially order these principles of justice as he did on the domestic level. Rather, he expects members of the society of peoples to make contextual judgments as to how to apply the principles in any given case, ranking some principles higher in priority than others depending on the circumstances. In the most extreme case, grave human rights violations by an outlaw state could justify military intervention, after the failure of negotiation and sanctions, to protect the victims' basic human rights (93–94n6).

In addition to deciding upon the law of peoples that guides the society of peoples in matters of international justice, the parties in the second-stage OP develop "guidelines for setting up cooperative organizations and agree to standards of fairness for trade as well as certain provisions for mutual assistance" (42). Rawls indicates that at least three international institutions would be established according to these guidelines: a fair-trade regulator, a cooperative banking system, and a confederation of nations like the United Nations, ideally conceived (42). These international institutions would enable the society of peoples to implement the law of peoples.

The second part of *LP* concerns the second-stage hypothetical social contract that extends membership in the society of peoples to decent hierarchical peoples. Such decent hierarchical peoples are not liberal or democratic, but they still meet minimum standards in consulting their people in political decision making and in protecting basic human rights. Decent peoples honor the right to life, liberty, personal property, and formal equality in matters of justice (65). Decent peoples are also nonimperial and nonaggressive toward other peoples. Rawls argues that hypothetical representatives of decent peoples would deliberate under the veil of ignorance in the second-stage OP and reach the same eight principles of the law of peoples to govern their interactions with other peoples.

Decent peoples would accept and implement these eight principles for the same reasons that liberal peoples do: the values of the law of peoples resonate with their own established political and moral values. Although they do not share the same domestic-level conceptions of justice, liberal and decent peoples have substantial overlap in their political values: respecting basic human rights; consulting the people in political decision making; and nonaggressive and nonimperial relations with other peoples. This overlap means that liberal and decent domestic-level conceptions of justice, although not identical, both resonate (in different ways) with the principles of the law of peoples. Because of this resonance, both liberal and decent peoples have reason to accept and put into practice the principles of the law of peoples as generated by the second-stage hypothetical social contract (69–70).

This pragmatic liberal attitude toward the fact of the pluralism of political systems is also shown in the third part of *LP*, in which Rawls outlines the relationship between the society of peoples and other kinds of nations. Outside the well-ordered liberal and decent society of peoples are three other types of nations, which are not well ordered. "Outlaw states" do not respect human rights and are typically expansive and aggressive. "Benevolent absolutisms" respect human rights and are not expansive and aggressive but do not maintain even a decent consultation hierarchy with their people (4). "Burdened societies" are not expansive and aggressive, but they "lack the political and cultural traditions, the human capital and knowhow, and often, the material and technological resources needed to be well-ordered" (106). While the ultimate goal of the society of peoples is to include all peoples and bring about a global peace through their voluntarily abiding by the law of peoples, the realization of this "ideal theory" depends on the long-term, complicated process of nonmembers' transformation of their basic structures such that they are either decent or liberal (57–58, 105).

As in his forerunner, Kant's 1795 essay "Perpetual Peace," Rawls does not expect this agreement to the law of peoples to happen anytime soon; it will be an extremely long-term process that first depends on the development of stable liberal and decent basic structures at the domestic level, and the subsequent development of international cooperative agreements and federations among such peoples. Moving beyond Kant, however, Rawls has a more expansive and tolerant conception of the membership of the society of peoples. Not only do liberal peoples (or republics, in Kant's theory) join the society, but decent peoples also join. By expanding the society of peoples to include nonliberal but decent peoples, Rawls shows his politically liberal and pragmatic toleration of a diversity of political systems.

Rawls's treatment of the relationship between the society of peoples and burdened societies is the most "controversial" aspect of his law of peoples

(*LP*, 37n43). As we have seen, the eighth principle of the law of peoples states that "peoples have a duty to assist other peoples living in unfavorable conditions that prevent them from having a just or decent social regime" (37). The controversy lies in determining the form of the assistance. As we have just noted, Rawls argues that among peoples that are not well ordered, "only burdened societies need help" (106). They are not necessarily poor, just as liberal and decent societies are not necessarily wealthy (106). In extreme cases of famine or natural disaster, they will need help from the society of peoples because they lack financial or other material resources. But even a disaster such as famine is generally brought about by "faults within the political and social structure, and its failure to instigate policies to remedy the effects of shortfalls in food production" (109). For this reason, even the duty to assist burdened peoples in times of famine is not entirely a matter of financial aid, but rather a matter of assisting burdened peoples with the longer-term project of reforming their political and economic systems such that they are fair with regard to planning internal food production and distribution. In general, however, burdened societies need help mainly because they lack the basic structure and "underlying religious and moral beliefs and culture" to "sustain a liberal or decent society" (106). The purpose of assistance to burdened peoples is to help them build a decent or liberal society in which human rights are respected and a decent consultation hierarchy, if not a democracy, is in place. "Merely dispensing funds," Rawls concludes, "will not suffice to rectify basic political and social injustices" (108). Rather, "an emphasis on human rights" in providing assistance to burdened peoples will help them, in the long run, voluntarily to adopt decent or liberal political and economic structures in a way that is consonant with the values of their own moral, religious, and political traditions (109).

Rawls on Justice for Women in Burdened Societies

Rawls exhibits a feminist turn in his political thought in theorizing the obligation of the society of peoples to protect women's human rights in burdened societies. Rawls uses the issue of women's human rights to explore how this "human rights" approach to the society of peoples' duty to assist burdened societies ought to work. He argues that the protection of women's human rights in burdened societies is a "decisive factor" in addressing some of the typical economic and political problems that drive burdened peoples' need for international assistance (*LP*, 109). Building on the work of economist Amartya Sen, Rawls argues that "population pressure" is an example of one such political and economic problem (109). Some states have addressed the issue of population pressure through gross violations of women's human rights; Rawls cites the case of China's "harsh restrictions on the size of families" (110), which encouraged the practice of female infanticide in that country. Other burdened peoples have addressed the issue of population pressure by delaying women's fertility through the advancement of their education and civil and political rights. Rawls cites the example of the Indian state of Kerala, which "in the late 1970s empowered women to vote and to participate in politics, to receive and use education, and to own and manage wealth and property. As a result, Kerala's birth rate fell below China's, without invoking the coercive powers of the state" (110). Through this contrast of China with Kerala, Rawls shows how attention to women's human rights in public policy and law helps to rectify broader conditions of economic and political inequality within such nations. In his words, "Burdened societies would do well to pay particular attention to the fundamental interests of women" (110).

This is a strong, yet potentially problematic, feminist statement for a political liberal. Rawls needs to explain how it is that we can expect burdened peoples to take into consideration "the fundamental interests of women" when their moral, religious, and political cultures may stand in tension with such interests as construed by the society of peoples. He begins by arguing that "the fact that a women's status is often founded on religion, or bears a close relation to religious views, is not in itself the cause of their subjection, since other causes are usually present" (110). Second, "all kinds of well-ordered societies" affirm human rights, including women's human rights—setting an international standard for justice for women (110). Third, no peoples can use religion to legitimate the subjection of women on the grounds that the subjection is necessary for the survival of the religion, just as no religion can justify the suppression of other religions on the grounds that its own survival depends on it (111). As a condition of offered assistance, it is thus reasonable for liberal and decent peoples to expect burdened societies to respect women's human rights.

The ability of the law of peoples to protect the rights of women in burdened societies depends heavily on the sensitivity of the society of peoples to the competing demands of tradition and democratization. If assistance to burdened societies encourages women's rights in a way that is seen as tolerant of traditional religious views and other comprehensive doctrines, it may eventually bring a shift in the voluntarily held values of the burdened peoples such that "democratic" or "feminist" values come to be seen as their own. This gradual shift in values and institutions may be slow in addressing women's rights, but the overlapping consensus about women's rights and social justice that develops within such nations will ultimately reinforce the stability of the reform.

As for extreme violations of human rights (including women's human rights), Rawls allows for severe political or economic sanctions or military intervention to stop them. As we have seen, the law of peoples states eight principles by which the society of peoples is meant to govern international relations among its members, as well as among peoples that are not well ordered. Although one of the eight principles is the principle of nonintervention, Rawls also includes a principle of protecting human rights (including women's human rights). Because he does not serially rank the eight principles, he is able to allow for liberal and decent peoples to deliberate about how best to apply the principles in particular contexts, and whether to make exceptions to the upholding of one principle in favor of defending another. In the case of extreme mass violations of basic human rights to life and security (i.e., genocide, politically induced famine), Rawls thinks it legitimate for liberal or decent nations to intervene militarily to protect human rights. Rawls believes that these are hard cases, especially when the outlaw state is nonimperial. He nonetheless allows liberal and decent peoples to wage a just, aggressive war against an outlaw state in order to protect basic human rights, once good-faith attempts at negotiation and sanctions have failed (93–94n6). In this way, the law of peoples is effective in stopping or deterring extreme violations of human rights because its allows for severe political and economic sanctions or emergency use of military force to prevent non-well-ordered peoples from, say, conducting systematic, genocidal rape and sexual torture of women as part of a civil war.[1]

Rawls posits as a basic principle of his theory of international justice that liberal and decent peoples—who meet at least a minimum threshold in respecting basic human rights and consulting the people in political decisions—ought to tolerate a diversity of religious and other comprehensive doctrines across and within nations that may sometimes stand in tension with their liberal or decent political values. For example, a liberal people (like Sweden) and a decent people (like Turkey) ought to tolerate the religious practice of polygamy among sub-Saharan burdened societies. Yet the society of peoples should use economic and political policies (such as United Nations conventions, international trade agreements, foreign aid tied to advancement of universal primary education, and support of a vibrant international nonprofit sector) to encourage such burdened peoples voluntarily to adopt a more robust and egalitarian conception of human rights. Once adopted, this ideal of human rights would eventually commit them to the voluntary outlawing of patriarchal practices such as polygamy and the adoption of at least a decent consultation hierarchy that includes "a majority of women" in "any group representing women's fundamental interests" (110). Because of his robust commitment to toleration of religious and other comprehensive doctrines, even when they stand in tension with liberal, democratic, or feminist values, Rawls provides a principled yet pragmatic framework for international justice that can be adapted to address most feminist issues in the long run, but with respect for the self-determination, religious and cultural diversity, and political pluralism of peoples.

Nussbaum's *Frontiers of Justice* as a Feminist Critique of Rawls's *Law of Peoples*

Frontiers of Justice is partly based on the arguments of Nussbaum's earlier book *Women and Human Development: The Capabilities Approach* (2000), which gives us her most extensive account of her theory of human capabilities and how it applies to issues of justice for women in developing countries. The capabilities approach sets forth a universalistic account of ten fundamental human abilities and the minimum threshold at which their real or potential voluntary realization characterizes a life of human dignity. In this later work, Nussbaum summarizes her capabilities approach and uses it to challenge the Rawlsian social contract approach to justice for developing nations. She uses examples of injustice against women in developing nations—such as how they "notoriously lag behind men in education, employment opportunities, and even in basic life chances"—to illustrate the limitations of Rawls's theory for promoting human rights in what he calls burdened societies (Nussbaum 2006, 225).

Nussbaum affirms several times that she is offering a kind of human rights approach to international justice that is in the same family of politically liberal conceptions of justice as Rawls's (2006, 80–81). Rawls and Nussbaum also build on the feminist liberal tradition of understanding women's rights as human rights. By defining women's rights as a kind of human rights, Rawls and Nussbaum mean that because women are human, they are entitled to the same rights as all human beings. They both cite John Stuart Mill's *Subjection of Women* (1869) as their inspiration for taking this philosophical position (IPRR, 156n58; Nussbaum 2000b, 53n39). In law and policy, this conception of rights is often described as "universal human rights," following the language and content of the UN's Universal Declaration of Human Rights (1948).

Both Rawls and Nussbaum support a nonmetaphysical conception of human rights. They argue that this nonmetaphysical conception of human rights does not depend on a universalistic theory of human nature or a metaphysical account of the person. For Rawls, this nonmetaphysical conception of human rights grows out of liberal and decent peoples' conceptions of justice; these human rights–oriented conceptions of justice are codified in law and practice over time, but they were initially products of modern European historical documents and political agreements that evolved in response to the sixteenth- and seventeenth-century wars of religion (PL, xxiii–xxviii). For Nussbaum, this nonmetaphysical conception of human rights grows out of a "world culture" of respect for the dignity of human life, which encompasses both ancient and modern, Western and non-Western political thought and practice (2006, 304). As a part of their broader commitment to political liberalism and its neutrality toward comprehensive doctrines, Rawls and Nussbaum take human rights as historical products of cultural and political traditions that we come to treat as universal moral principles that ought to govern domestic and international politics. Rawls and Nussbaum succeed in different ways in defending this nonmetaphysical view of human rights alongside their commitments to promoting justice for women in developing countries—a point I explore at length below.

Although Rawls and Nussbaum describe human rights differently, there is basic overlap in their conceptions. Rawls provides a general list of basic human rights, while Nussbaum offers a more detailed and expansive account of human rights in her list of the ten human capabilities. For Rawls, human rights are "the right to life (to the means of subsistence and security); to liberty (to freedom from slavery, serfdom, and forced occupation, and to a sufficient measure of liberty of conscience to ensure freedom of religion and thought); to property (personal property); and to formal equality as expressed by the rules of natural justice (that is, similar cases be treated similarly)" (LP, 65). For Nussbaum, the capabilities (which she says can be understood as rights in a political sense) come in the form of a more detailed list than Rawls's basic human rights. Summed up, the capabilities are life; bodily health; bodily integrity; senses, imagination, and thought; emotions; practical reason; affiliation; living with concern for other species and nature; play; and control over one's environment (2006, 76–78). Indeed, we could read Nussbaum's ten capabilities as an expanded and more detailed version of Rawls's list, and Rawls's list as implicitly endorsing much of what Nussbaum spells out. But even under this interpretation, Nussbaum specifies some rights that go beyond what Rawls implied, including the right to play, the right to live with concern for nonhuman animals, and the right to emotions.

However simple and elegant it is as an abstract philosophical account of universal human rights, Rawls's basic human rights approach reveals its limits at the level of policy for women. As Nussbaum indicates, a more detailed list of human capabilities aids the crafting of policies and laws that deal with women's particular challenges in holding or exercising human rights that are typically compromised by their status as females. Her detailed account of the human capabilities of bodily health and bodily integrity, for example, includes rights to "reproductive health," "opportunities for sexual satisfaction," "choice in matters of reproduction," and security "against sexual assault and domestic violence" (2006, 76). While Nussbaum conceives of these as human rights that apply to both women and men, she focuses on how they apply to women because of existing conditions of economic, political, social, and sex-based inequality for women, especially in the developing world.

For example, the fact that women are more often victims of rape and domestic violence than men are makes women's human right to be free from these forms of violence even more imperative in the advocacy, education, policy, and law-making realms.[2] Nussbaum's delineation of such human rights in her ten capabilities situates rape and domestic violence, which have often been marginalized as "women's issues," at the center of human rights advocacy. Her capabilities approach thus challenges us to treat rape and other forms of sexual or sex-based violence against women as grave human rights violations in the policy and legal realms. While Rawls's broader and simpler list can be interpreted to include such specific applications of the basic human rights, Nussbaum's list is stronger from a feminist theoretical and policymaking perspective because it highlights particular areas in which women's human rights need addressing because of the existing inequalities between the sexes.

Beyond her feminist criticism of Rawls's "thin menu" of human rights, Nussbaum questions several key assumptions that guide Rawls's social contract approach to international justice (2006, 243, 248). These assumptions include the human rights orientation of the representatives in the second-stage OP; the free, equal, and mutually advantageous status of the parties in the international agreement to abide by the law of peoples; and the value of the two-step social contract process. In critiquing these assumptions, Nussbaum contends that justice for women, especially justice for women in developing countries, is hard to achieve under the social contract approach of *LP*. While this may be true in certain cases, Nussbaum's critique of Rawlsian social contract theory misrepresents some aspects of Rawls's approach and thus misses some of its usefulness for women's rights issues.

First, Nussbaum argues that the representatives of the nations in the second-stage OP would not be able to represent the human rights of their peoples "taken as a whole" because "large segments of the population (women, racial minorities) may be completely excluded from governance" (2006, 223). Initially, Nussbaum's critique can be seen as applying only to the extension of the law of peoples to decent peoples, since liberal peoples, by definition, abide by "justice as fairness" and its stringent demands for equal civil and political rights and fair equality of opportunity, regardless of gender, race, or other identity traits (*LP*, 50). Upon reflection, however, the critique seems misplaced even for decent peoples. Under Rawls's definition, decent peoples respect women's human rights, in part by ensuring a majority representation of women in decision-making bodies concerning women's fundamental interests (110). Theoretically, women's interests would be taken into account in the second-stage social contract, because their interests are taken into account in the basic structures of liberal and decent societies. It is important to recall that Rawls theorizes the development of the society of peoples as an extremely long-term process, because it takes a lot of time, mistakes, and reforms for peoples to develop the liberal and decent basic structures that would respect human rights in the rigorous way that his theory of international justice demands. For example, Rawls's stringent definition of a liberal basic structure as including universal health care would certainly make today's United States a decent people, at best (50). In other words, Rawls isn't thinking of peoples as they are now but as they should and could be. In the long run, Rawls hopes that there will be a substantial number of liberal and decent peoples who are inclined to abide by the conception of universal human rights that lies at the heart of the law of peoples.

Second, Nussbaum contends that the social contract idea demands that the nations that accept the principles of the law of peoples must mirror the representatives in the hypothetical OP in their freedom and equality (2006, 250). Given Rawls's insistence that the OP, at both levels, is hypothetical, this critique seems misguided in its literalism. The principles of justice that are generated from this hypothetical decision-making procedure are principles that we can imagine such "free and equal" rational parties agreeing to behind a veil of ignorance. The principles are not actually agreed upon by similar people in real societies, domestic or international. The law of peoples is adopted by real people because (1) it is seen as emerging from a philosophical thought experiment that is fair in procedure and outcome, (2) it resonates with conceptions of justice in a liberal or decent society, and (3) it overlaps with their moral values in a way that encourages them to follow the principles out of duty. At most, Rawls's theory presumes that nations that accept the principles of the second-stage social contract are "free and equal" *enough*[3] to enter into an agreement with others to follow the conception of justice produced by the thought experiment. At the international level, this could mean liberal and decent peoples of differing wealth and power entering into "regional associations or federations of some kind, such as the European Community," but understanding the fundamental terms of their joining to be mutual respect of members' equal rights. By calling the members of the society of peoples "free and equal," Rawls means only that they expect to be treated as equals in freedom (autonomous peoples with equal rights compatible with those of other peoples) by their fellow members. He explicitly states that they are not equals in wealth, political systems, cultures, or any other respect (*LP*, 115); indeed, he envisions a reasonable pluralism characterizing the members of the society of peoples.

Nussbaum similarly argues that Rawls's use of the social contract idea develops a theory of international justice that requires "mutual advantage" as a condition of joining the society of peoples (2006, 228, 249). She problematizes the goal of mutual advantage by citing the radical economic differences between nations today; wouldn't less well-off peoples thus stand to gain far more from entering the society of peoples than better-off peoples? For example, Sierra Leone, with a per capita GDP of $470, would stand to benefit more from an international social contract than would the United States, with a per capita GDP of $34,320 (224).

While "mutual advantage" is one of the features of decisions made by the parties within the OP at both the first and second stages, it is important to recall that the characteristics of the OP's structure and outcome (at both levels) are hypothetical. The properties of the rational parties, and the properties of their decision-making scenario, are not "real" traits of "real" people or peoples. They are merely the abstract, philosophical conditions of a fair procedure for generating a fair conception of justice.

In the real world, liberal and decent peoples would use the second-stage social contract thought experiment to grasp the principles of the law of peoples that ought to govern their international relations. It is a second step for them to agree to put into practice those principles by establishing an international federation, bank, or trade agreements. The members of the federation of the society of peoples need not be wealthy, but they need to be decent or liberal in their basic structures. A liberal or decent but poor Sierra Leone could join the federation of the society of peoples. Although it stands to gain more financially from its membership than a wealthy decent or liberal people does, there remains a fundamental, fully mutual advantage of the federation: the realization of equal respect and equal rights among members such that self-determination, human rights, and peace prevail among and within them. It is this basic political level of mutual advantage that Rawls expects for members of the federation of the society of peoples, not an equal advantage in every respect (finances, security, resources, etc.). Just as members of the federation of the society of peoples must be "free and equal" *enough* to join such an organization and abide by its principles, the federation must be "mutually advantageous" *enough* to protect its members' equal rights.

So, while the conditions of the OP are those of mutual advantage and rough equality of the free, rational parties, these conditions do not perfectly mirror the real world for Rawls. Equality (such as equal protection of basic human rights) is something that has to be "made," through inculcation of liberal and decent values in families, education, legislation, and judicial decisions. Moreover, Rawls's goal is never to create an across-the-board equality, but specifically an equality of rights, at both the domestic and international levels. In the international realm, Rawls accepts big economic and political differences between peoples as part of the moral landscape, even after the federation of the society of peoples is established. The point of the law of peoples is not to eradicate difference within and among peoples but to adjudicate difference fairly across peoples in a way that promotes their self-governance.

Third, Nussbaum questions the value of the two-step social contract process Rawls outlines in *LP*. In particular, she argues that Rawls assumes the "fixity" of the domestic structure of nations (2006, 234). While Rawls certainly argues that peoples must reach either liberal or decent status prior to entering into the second-stage social contract, he never claims that this status is fixed, once reached. Many of his historical examples suggest an acute awareness of how peoples can shift across categories—Germany, for example, has arguably been an outlaw state, a burdened society, a decent people, and a liberal people at different times in the past century.

Despite these flaws in her representation of Rawls's use of social contract theory, Nussbaum successfully uses the example of CEDAW to explain how his theory of international justice could be insufficient in defending women's human rights (2006, 243). The Convention on the Elimination of

All Forms of Discrimination Against Women (CEDAW) was adopted by the United Nations in 1979. It is now treated as a universal code for protecting women's human rights, especially by developing nations. First, Nussbaum notes that CEDAW's broader and deeper sense of the rights of women, and the demands for legal reform that it places on its ratifying members, stand in tension with Rawls's basic human rights approach. For example, CEDAW has provisions concerning the outlawing of marital rape that require major reforms in domestic laws in many nations. Nussbaum argues that the society of peoples would not require that its members reform their laws to fit the more rigorous and expansive human rights expectations of international treaties like CEDAW that bear on members' "domestic arrangements" (243).

While Rawls certainly endorses the authority of international agreements within his conception of the society of peoples, an agreement like CEDAW still may not promote justice for women within Rawls's theory of international justice (*LP*, 42). For one thing, the traditional public-private distinction in liberal peoples masks the ways in which women are oppressed in domestic-level familial and religious structures (Nussbaum 2006, 321–22). A society may look liberal but still permit marital rape, as the United States did until it began to pass laws against it in the late 1970s (Hines and Malley-Morrison 2005, 42). For another thing, enforcing a feminist international agreement such as CEDAW is even more difficult in burdened societies, which stand outside of the norms of the law of peoples. Nations without democracy or decent consultation hierarchies, without laws and institutions that support women's rights, or without universal primary education appear to be unlikely implementers of CEDAW. Even nations, like Uganda, that have made great strides in implementing the principles of CEDAW in their laws and political institutions may only pay lip service to religiously controversial aspects of the policy, such as phasing out polygamy (Shore 2010). According to Nussbaum, Rawls's focus on nationality and retention of the public-private distinction makes his theory of international justice of questionable use in addressing these deeper issues of women's human rights.

In Conflict and Peace: Defending Women's Human Rights in the Developing World

While Nussbaum provides a detailed critique of Rawls's *LP* from a feminist perspective, she does not view herself as departing from Rawls's political liberalism so much as exploring and addressing some of the limitations of his theory. I follow Nussbaum in viewing her theory as broadly compatible with Rawls's. Their "human rights" approaches to international justice can be applied alongside each other in different cases that require different political emphases. Rawls's stricter conception of political liberalism—which derives from the principles of nonintervention and national self-determination—allows his theory to generate long-term solutions to problems of women's oppression in burdened societies characterized by religious or violent conflict. The universalistic capabilities language and more specific menu of women's human rights found in Nussbaum's approach make it better for addressing secular problems of women's oppression in more peaceful burdened societies.

Nussbaum's capabilities approach gives a principled yet practical standard with which policymakers can judge a richly human quality of life for women across nations. This universalistic standard of human development is particularly useful for advocating and institutionalizing the feminist reform of secular institutions in developing nations. Mobilizing women to vote or run for office, enabling or encouraging women to own property, or promoting microfinance loans for female business owners can be examples of fairly uncontroversial secular political issues that the capabilities approach can address in its universalistic terms without seeming culturally insensitive or imperial to a developing nation. Even the more contested issues of domestic violence and marital rape are better addressed by the capabilities approach, which enshrines the rights to bodily integrity and bodily health at the core of a richly human life. In a context of religious conflict, however, Nussbaum's capabilities approach fares less well than Rawls's in defending more contested human rights. Women's human right not to be raped in marriage is contested by many patriarchal religious traditions, familial structures, and legal systems in the developing world. Despite its heritage in the tradition of political liberalism, the universalistic language of the capabilities approach, coupled with its establishment of a global minimum standard of human flourishing and human rights for men and women alike, makes it less likely to garner the support of nonliberal or fundamentalist religious peoples whose traditions demand that they see and treat women as unequal or subordinate to men.

LP offers a more useful framework than Nussbaum's for navigating issues of international justice for women in situations of religious or violent conflict. The broad outline of the law of peoples, with its flexible framework for prioritizing its principles according to context, makes it more adaptable to situations of severe conflict. In a case of religiously driven conflict over women's right to be free from polygamy or marital rape in a burdened society, its respect for the autonomy and religious diversity of nations will not permit the law of peoples to intervene directly. On the other hand, the law of peoples can justify giving foreign aid on the condition that the recipient make advances in less controversial, secular areas of human rights, such as universal primary education. The advancement of girls and women in society through education will eventually make it more likely that the nation will voluntarily outlaw polygamy, marital rape, and other controversial patriarchal practices inflected by religion.

In situations of violent conflict that involve gross, systematic human rights violations (such as genocidal rape as a weapon of war in Rwanda or the Congo), the law of peoples can justify military intervention on behalf of the victims. This exception to the principle of nonintervention is not the norm for Rawls's law of peoples, however. The stricter political liberalism of the law of peoples makes it tend toward tolerance of big cultural differences between and within nations, across and beyond the society of peoples. But as a species of the human rights approach to international justice, Rawls's law of peoples will never tolerate the intolerable. For this reason, feminist liberals should explore Rawls's theory of international justice as a resource for defending women's human rights in the hardest cases of religious and violent conflict.

Notes

1. I follow Catharine MacKinnon in defining rape as genocidal when its aim is the "destruction of peoples as such." In MacKinnon's words, "Each act of sexual abuse committed with an intent to destroy the (usually) women of the group defined by its nationality, ethnicity, religion, and/or race is therefore legally an act of genocide" (2006, 221).

2. One-third of female humans are victims of sexual or domestic violence over the course of their lives (UN Commission on the Status of Women 2000).

3. I credit Nathaniel Lee for this turn of phrase.

7

Jean Hampton's Reworking of Rawls

Is "Feminist Contractarianism" Useful for Feminism?

Janice Richardson

In this chapter, I look at Jean Hampton's "feminist contractarianism" to explore what she means by the term "contractarian" and to evaluate the extent to which her work can provide resources for feminism. I start by considering Hampton's proposed contract in order to draw out some critical distinctions between her Kantianism and that of Rawls. Crucially, Hampton keeps the image of what it means to respect *oneself* as a free and equal person in play, as providing a test that should be repeated whenever anyone wants to decide if a particular relationship is just. This is in contrast with Rawls, who employs the image of free and equal persons making a contract as part of a single thought experiment from which he derives his principles of justice. In Rawls's outline of the OP, he employs the image of free and equal persons reaching an agreement with the following purpose:

"to make vivid to ourselves the restrictions that it seems reasonable to impose on arguments for principles of justice and therefore on these principles themselves" (*TJ*, 18). Rawls does not seek to repeat this procedure once he has derived the principles of justice, whereas Hampton uses the image of free and equal persons to prompt us to think about justice in everyday relationships. I therefore position Hampton's work alongside that of other feminist thinkers who are concerned with the implications of everyday beliefs, especially beliefs regarding morality itself, and their broader psychological effects—in particular the question of shame. In *TJ*, Rawls views self-respect (which, at this point, he does not differentiate from self-esteem) as "perhaps the most important primary good" (440). Hampton's analysis highlights the importance of "self-worth," which she views in Kantian terms, such that all persons have equal self-worth, despite the fact that they can (wrongly) come to view themselves as subordinate. Finally, I look at feminist critiques of contractarianism, and of the use of tests that refer to personhood, to evaluate the extent to which Hampton's work on "feminist contractarianism" can be read so as to avoid these major criticisms.

Hampton's Contract

Unusually in analytic philosophy, Hampton appeals to the importance of the imagination in arguments. She refers to "contractarianism" as possessing a number of diverse strands of thought, which have in common the fact that they employ a "picture" or "image" of free and equal persons (Hampton 2007a, 8). I examine Hampton's proposed test for fairness in relationships below. Significantly, her appeal to a hypothetical contract performs only one philosophical job. It focuses attention upon the idea of free and equal persons, or, to put this another way, on the wrongness of subordination in its nuanced forms, including those expressed and perpetuated through moral beliefs themselves. Rather than derive principles of justice from her hypothetical contract, Hampton provides a Kantian framework through which these specific instances of injustice can be understood and recognized. In particular, she shifts our attention from the public sphere to highlight injustice that occurs in intimate relationships, thereby also emphasizing the need for justice within such relationships.

In "Feminist Contractarianism," Hampton considers the responses of two eleven-year-old children to moral questions asked in Carol Gilligan's famous study. When asked the question "When responsibility to oneself and responsibility to others conflict, how should one choose?" Jake replies, "You go about one-fourth to the others and three-fourths to yourself" (Gilligan 1982, 35–36, quoted in Hampton 2007a, 3). In contrast, Amy responds,

> Well, it really depends on the situation. If you have a responsibility with somebody else [sic] then you should keep it to a certain extent, but to the extent that it is really going to hurt you or stop you from doing something that you really, really want, then I think maybe you should put yourself first. But if it is your responsibility to somebody really close to you, you've just got to decide in that situation which is more important, yourself or that person, and like I said, it really depends on what kind of person you are and how you feel about the other person or persons involved. (Gilligan 1982, 36, quoted in Hampton 2007a, 3)

One can see clearly whose wishes would be prioritized if Jake and Amy were to marry. Hampton views both children as producing immature answers that reflect the position of one who is learning to dominate and one who is learning to be subordinate, respectively. She points out that we are used to viewing selfishness as a moral problem, but not the question of selflessness. This may be either because men are treated as the norm or simply because we appreciate that selflessness is not so antisocial and, more worryingly, can be useful for others who may take advantage of it.

With this in mind, Hampton produces a test for fairness within relationships: "Given the fact that we are in this relationship, could both of us reasonably accept the distribution of costs and benefits (that is, the costs and benefits that are not themselves side effects of any affective or duty-based tie between us) if it were the subject of an informed, unforced agreement in which we think of ourselves as motivated solely by self-interest?" (2007a, 21). Hampton is not assuming that everyone is motivated merely by self-interest or that emotions are irrelevant, points that I discuss below. But the aim of her test is to draw attention to unfairness in intimate relationships, particularly in situations where a woman assumes that she is "a different type of person" for whom morality involves always putting others' interests first.

The role of the contract for Hampton can be examined by considering David Gauthier's rejection of her view that "every contract theory . . . has used the idea of a contract as a heuristic tool that points us toward the correct form of moral reasoning and has not relied on the notion of contract in any literal way to do any justificatory work" (Hampton, quoted in Gauthier 2007, xii).[1] Gauthier sees this argument as "selling contractarianism short." He claims that the contract itself does do more work: "Only by determining what rational persons would agree to in a suitable pre-moral situation can we give content to and a rationale for moral principles. Proposed or alleged moral principles can be put to the contractarian test—might they be agreed to by rational persons seeking principles to govern their interactions? I leave to the reader the question whether this role is merely a 'heuristic tool'" (xii).

Gauthier's disagreement illustrates his fundamental difference from Hampton. As Hampton makes plain, the image of a contract, for her, is a heuristic device that allows us to focus upon the idea of treating each other as "free and equal persons." This device is then applied to test fairness in specific relationships. As in Rawls, there is no appeal to an actual, legally binding contract, of course. While Gauthier has argued that the contract thought experiment does more work, I have argued elsewhere that this is because he is concerned with game theory—derived from his Hobbesian position, in which one's worth is simply one's value or price (Richardson 2009, 28). Thus, for Gauthier, one's worth can vary according to one's abilities at different points in life. In contrast, the central point of Hampton's work is that we are all to be treated as persons with equal intrinsic worth—a position she describes as Kantian, irrespective of Kant's actual views on women and race (for example, Kant 1991b, 76–116). To summarize, her appeal to a contract is only used to draw out an image of free and equal persons (i.e., to rule out subordination). In contrast, Gauthier envisions self-interested individuals employing game theory to pursue their interests, thereby making the contract do more work as the mechanism through which game theory is understood.

Carole Pateman (2009), in a review of Hampton's work, argues against the need for the device of contract at all. This is not a point that Hampton would resist, given that she is clear that contract is only doing the work of forcing us to think of *ourselves* (as well as of others) as free and equal persons. In doing so, she asks what we would be willing to agree to if we were to consider an intimate relationship in terms of self-interest, that is, if neither our emotions nor our sense of duty were used as leverage. This inserts a Hobbesian element into her work that needs to be clarified.

Hampton explains that there is one element of Hobbes that she finds evocative: the phrase "we are not under any obligation to make ourselves prey to others" (Hampton 1999, 236).[2] Hampton's central ("Kantian") point (that respect for equal persons includes respect for oneself as well as others) overlaps with Hobbesian self-interest in that both provide (different) criticisms of self-sacrifice. For Hobbes, it is irrational not to be self-interested, whereas Hampton's reading of Kant provides a criticism of behavior in which anyone treats herself as less worthy of respect than others. Thus Hampton remains true to the Kantian view of persons, while employing the Hobbesian element in a test that is designed to highlight self-sacrifice. She shares Gauthier's concern with exploitation, which she describes as occurring if "one party relies upon the affection or duty felt by another party to use that other party to her detriment" (2007a, 20). Affection and duty are thus excluded from the test, discussed further below. I also return below to the question of exploitation and its relationship to subordination.

Gauthier recognizes the feminist issue, and Hampton quotes from him approvingly: "[Our sociability] becomes a source of exploitation if it induces persons to acquiesce in institutions and practices that but for their fellow-feeling would be costly to them. Feminist thought has surely made this, perhaps the core form of human exploitation, clear to us. Thus the contractarian insists that a society could not command the willing allegiance of a rational person if, without appealing to her feelings for others, it afforded her no expectation of net benefit" (Gauthier 1987, 11, quoted in Hampton 2007a, 20). Hampton's test may appear to be at odds with Kant's moral position, in which duty and emotion are opposed. Rather than view duty as superior to emotion, Hampton treats both as factors that must be excluded when deciding what burdens and benefits could be reasonably accepted within a relationship. She appears to be stating that women may be exploited both because of their emotions (or fellow feeling, as Gauthier puts it in the passage quoted above) and because of their duty to others. However, it is clear that she remains Kantian. Without stating it in such terms, her test distinguishes between "true duty" (which would be to treat everyone, including oneself, as free and equal) and "a wrongly perceived duty," which (like emotion) is employed as leverage in an unjust relationship. In other words, she is using the idea of the equal worth of persons in order to attack other "moral" positions that perpetuate relations of subordination.

Ruth Sample (2003, 117) asks why ties of affection should be excluded from Hampton's test. The answer lies in Hampton's concern about the use of both emotion and duty as leverage. One reason that Hampton gives is that emotions cannot be distributed. I would argue that it is also the case that if one party is taking more benefits over a period of time, as an integral part of the relationship, then it is unacceptable for him (or her) to try to justify this by saying, "you do more for me because you love me (or our family) more than I love you," or "as a woman, it is your duty to do more." Hampton's belief, which I share, is that without justice there is unlikely to be respect (either self-respect or respect by the other party), which is a prerequisite for love. By keeping this "contract test" in play, rather than using it, as Rawls does, as the basis of a one-time procedure for deriving principles of justice, Hampton forces us to analyze daily acts of subordination.

In the next section I consider this issue of self-worth, along with self-respect and self-esteem, in more detail. I agree with Elizabeth Brake's argument in chapter 3 that a view of self-respect (a recognition of one's equal self-worth) as a primary good provides resources for feminism. For Brake, it forms the basis of a critique of family practices that undermine children's self-respect. My argument is not focused upon children, but I agree that if anyone in the household is treated as a subordinate (a position traditionally associated with wives), then this will teach children a lesson about inequality that is morally problematic. As we saw in Gilligan's study, it has the potential to undermine Amy's self-respect and to teach Jake to behave unjustly. As Brake observes, this reveals a tension within Rawlsian liberalism itself.

Self-Worth, Self-Respect, Self-Esteem, and Shame

As noted above, Rawls states that self-respect is "perhaps the most important primary good" (*TJ*, 440). It is clear from his definition, however, that he sometimes has in mind self-esteem, which (as he acknowledges) he is treating on those occasions as synonymous with self-respect. Rawls draws upon psychological arguments to describe self-respect as dependent upon both having a worthwhile plan of life and believing that one has the ability to carry it out. The basis of self-respect is thus described as "(1) having a rational plan of life, and in particular one that satisfies the Aristotelian principle,[3] and (2) having our person and deeds appreciated and confirmed by others who are likewise esteemed and their association enjoyed" (*TJ*, 440).

As David Sachs (1999) makes clear, self-respect and self-esteem differ. This is illustrated by the fact that it appears to be an improper use of language to say that someone has too much self-respect, whereas it is clear what is meant when someone is accused of having too much self-esteem. Loss of self-respect carries the more serious connotation of belittling or degrading oneself (or of being treated as less than a free and equal person, in Kantian terms). In contrast, loss of self-esteem may simply mean bringing one's inflated opinion of a particular ability into line with others' estimation of it. After suitable criticism, I can come to terms with my inability to play the cornet without its damaging my sense of worth as a person, or feeling that it gives anyone the right to treat me as a subordinate as a result. This distinction is compelling, and Rawls accepted it in his later work (*PL*, 404n39).

Sachs does not indicate whether Rawls should be understood to be talking about self-worth or self-esteem in *TJ*. My view is that the two concepts need to be unpacked, as some of Rawls's arguments address the importance of self-respect when in fact they describe self-esteem. So, for example, the passage quoted above, on one's plan of life and the need for the appreciation of others, describes the basis of self-esteem (Thomas 1999). Our self-esteem is based upon our estimate of our abilities—taking into account feedback from others—and can vary over time. Self-respect, by contrast, is based upon the view that we have equal self-worth as persons, that is, that we are not to be treated as subordinate, and is clearly of primary importance to us. Self-respect cannot be treated as a good that can increase or decrease, because it is implicit in what it means to be a free and equal person—as discussed in the OP and attributed to Kant (Hampton 1980). Henry Shue (1999, 192) argues that Rawls did not intend that self-respect would ever be unequal within the difference principle, because it is never the case that unequal self-respect is to the advantage of those with the least self-respect.

Given that Rawls draws upon psychological literature to consider the problems of shame and guilt (*TJ*, 443; and see esp. Piers and Singer 1953), it is useful to consider the changes that have been made to this literature in the past forty years. There have been arguments that the West has moved from being a guilt culture and has become a shame culture (Giddens 1991, 69). Leys (2007, 6) also points to the shift from ideas of "survivor guilt," which was viewed as integral to posttraumatic stress disorder by the American Psychiatric Association's *Diagnostic and Statistical Manual of Mental Disorders* in 1980 but has since been downgraded, to a greater focus upon shame. Tangney and Dearing (2002) employ empirical research to produce a definition of shame that is close to the classic approach of Piers and Singer discussed by Rawls (*TJ*, 443). However, they reach a slightly different conclusion regarding the extent to which shame is a "moral emotion," that is, an emotion likely to foster moral behavior. They produce a number of key factors that distinguish shame from guilt. Experiences described as shameful involve greater distress and a devaluation of oneself as a whole person, unlike experiences linked with guilt, in which one exhibits concern that a particular action or omission may have harmed another and that a moral or legal code has been breached. Those who feel guilt will thus try to make amends in a direct way, with an apology, for example. Therefore, Tangney and Dearing argue, guilt is a useful emotion that facilitates living together. In contrast, shame prompts one to hide from others. Whereas guilt is not associated with anger, shame can involve anger and blaming others for one's perceived deficiencies and sense of inferiority.

It is worth considering shame in detail to illustrate the narrowness of Rawls's conception of it and to explain why Hampton's feminist work is useful in this area. For Rawls, individuals experience shame because either their plan of life has been shown to be shoddy or they view themselves as unable to complete their plans successfully. John Deigh (1999) argues that Rawls's account of the causes of shame does not really fit with many situations in which individuals report experiencing shame in our society. He points out that these include situations in which the experience of shame is associated with being a member of a particular social group. He gives the examples of (1) an aristocrat who feels shame for behaving like a plebeian, and (2) a Native American who describes feeling shame when he is not able to participate in a war dance, having been uninterested in his family history.

I agree with Deigh that Rawls does not account for these experiences of shame, but I find his examples slightly odd. With regard to race, it does not touch upon the pain of persons who have been stigmatized. This can be compared with a moving example given by Moira Gatens in her analysis of privacy. Gatens describes a woman's confusion when her grandmother, over the course of many years, urged her not to invite anyone to the house and was very anxious when anyone official visited. It later transpired that her grandmother was classified as "black" and had been passing as white within a racist society. This secret caused her to feel shame—not because of a chosen life plan or an inability to carry it out, but because of the stigma cruelly attached to her racial classification (Gatens 2004, 125–29).

Hampton's analysis of morality is similarly compelling because she is alert to the mundane (sometimes petty, yet vicious) treatment of individuals as subordinate. Hampton also takes care to differentiate between those who view themselves as superior—as in Deigh's aristocrat example—and those who view themselves as inferior. For example, Hampton argues that the impact of serious crime can affect victims in different ways. They may feel demeaned or, worse, "diminished" by the crime. Recall that Hampton's starting point is that we are free and equal persons and hence have equal worth. Self-worth is the basis of self-respect, and, unlike self-esteem, it does not fluctuate depending upon our value to others. The criminal has (wrongly and falsely) held himself out as being of greater worth, as able to use the victim as a means to an end, rather than as an equal person and thus worthy of respect. Therefore, the victim rightly feels demeaned by the moral injury. Hampton therefore defines moral injury as "damage to the realization of a victim's value, or damage to the acknowledgement of the victim's value, accomplished through behavior whose meaning is such that the victim is diminished in value" (2007c, 127). She also recognizes that, as in Deigh's example of the aristocrat, someone who has not been accorded the deference that he believes is due to someone of his class, status, or gender will also feel demeaned: "A white person who is forced to sit next to a black person on a bus might believe this demeans her by making it appear that they are of equal rank and value and should thus be accorded equal treatment" (1988a, 49). But in this instance, Hampton argues, she is wrong (both morally and epistemologically) to feel demeaned, because her equal worth as a person was not under attack when special privileges were denied to her.[4]

Hampton also describes how the victim of a serious crime can come to feel "diminished," an even more insidious process. This is the term she employs for a situation in which a victim does not feel demeaned, because she views herself as the type of person who deserves nothing better and thus naturally has to put up with this sort of treatment by others. This attitude demonstrates that the victim views herself as subordinate, which Hampton describes as a problematic "theory of worth": "Alternatively, one's theory [of worth] can prevent one from feeling demeaned by treatment which seems, given *our* view of her value, demeaning. A rape counselor once told me of a woman who failed to tell anyone that she had been raped by a man she knew because she thought this was the sort of thing women had to 'take' from men" (1988a, 49). It may be that the victim felt this way prior to the crime, or that the crime itself taught her this (objectively wrong) lesson. She thus lost her sense of self-worth and was diminished as a result of the crime. For Hampton, given the axiom that we are free and equal persons with (fixed) equal self-worth, this perception is objectively wrong: "By victimizing me, the wrongdoer has declared himself elevated with respect to me, acting as a superior who is permitted to use me for his purposes. *A false moral claim has been made*" (Hampton 1988b, 125, emphasis added).

In his foreword to *The Intrinsic Worth of Persons*, quoted above, Gauthier asks, if we are not actually degraded, then how can we appear as such (2007, xiii)? He does not explore this question very deeply, although, unlike Hampton, he uses Hobbes as the source of his arguments regarding morality (Gauthier 1987). It is useful to consider what happens when Gauthier's claim is reversed. If our worth is simply to be viewed as our value to others (as Hobbes argues), then we cannot be viewed as being degraded by an action that fails to respect us. We are simply being treated in accordance with our current worth. To be degraded in Gauthier's Hobbesian terms thus refers simply to the idea of sliding down a ranking chart, which is accurate, based upon others' judgment of our "market position." For Hampton, by contrast, the experience of being degraded is based upon a comparison between how one is actually treated and how one ought to be treated. Hampton's potential response to Gauthier's question highlights the point that, even when treated badly, the victim is still a person and so does not actually lose the right to be treated with respect, even if she is so ground down that she mistakenly thinks that she does.

This Kantian move has implications for Hampton's approach to intimate relationships. Hobbesians like Gauthier will take the best deal on offer. A woman who wants a husband and children may also prefer not to be treated like a drudge, but if she must accept being treated like a drudge in order to have a family, then she will accept it. From Hampton's "Kantian" position (which is, as noted above, some distance from Kant's actual view of women), it could be argued that some options should be refused. In other words, there is a problem with a relationship that involves daily activities that undermine self-respect. This raises radical issues with regard to intimate relationships generally and, in particular, those based upon the model of traditional marriage.

While the common, everyday treatment of women in families may not be nearly as severe as crimes against women like rape and assault, the basis of Hampton's analysis is clear throughout. She explores the subtleties of her view of self-worth in her papers "The Wisdom of the Egoist" (1997) and "Selflessness and the Loss of Self" (1993). Egoists illustrate the meaning of self-worth because their sense of self-worth does not alter depending upon their value to others. The fault of the egoist is that this insight is not applied to his view of others. In "The Wisdom of the Egoist," without reference to the centrality of self-respect in Rawls, Hampton considers how difficult it is for some people to maintain a sense of equal worth when treated as subordinate by others, compared with those who are educated to believe in their own entitlement to have their voices heard above all others.

This concern with equal worth also informs Hampton's work on forgiveness and public justice. Someone should not be forgiven if it appears that she has not moved from the belief that it is acceptable to treat another person as less worthy of respect (Hampton 1988b, 147–57). Similarly, Hampton is Kantian with regard to the demands of public justice, arguing that this should send the message that the criminal is not right to hold himself out as more worthy than others, such that he is able to treat the victim as a means to an end, that is, as subordinate. Where this has not been done—for example, in cultures where rape is not properly prosecuted or where racist crimes go unpunished or are punished inadequately—the law fails to send this message of the equal self-worth of all citizens: "To vindicate the victim, a retributive response must strive first to re-establish the acknowledgement of the victim's worth damaged by the wrongdoing, and second, to repair the damage done to the victim's ability to realize her value" (Hampton 2007c, 135).

It is useful to consider rape in this context in order to draw Hampton's arguments together. This crime is often associated with feelings of shame and of being unclean on the part of the survivor. In Hampton's analysis, the survivor is right to feel demeaned, in that she has been treated as less than a person, but this does not affect her objective self-worth. As discussed above, if she views this treatment as her natural lot in life, then, in Hampton's terms, she is "diminished" and wrongly views herself as less worthy than others of respect as a person, which is associated with shame. Hampton describes the attitude of men who treat women as objects to be used whenever they want, commenting, "If he succeeds in raping her, not only is this the message one 'reads off of' the rape, more distressingly, the rape is a kind of event which seeks to make that diminished status a reality. She has been used as though she is an object, so she is one, right?" (2007c, 131). Thus shame, in this case, is not linked with any view of being shown to have a shoddy life plan or of an inability to fulfill it, as Rawls indicates. It is associated with an identity—that of being female—the subordination of which is "demonstrated" to her by the rape, as a type of hate crime. The rape effectively says, "you are a lesser type of person, and I can treat you in this way." For Hampton, the objective fact that this is not true should be demonstrated publicly: "a decision not to punish wrongdoers such as the rapist is also expressive: it communicates to the victim and to the wider society the idea that such treatment, and the status it attributes to the victim, are appropriate, and thus, in the case of the rape victim, reinforces the idea that women are objects to be possessed and are 'there for the taking'" (2007c, 133). Even where this has appeared true in some cultures, in that rape has gone unpunished, it is important to keep hold of Hampton's axiom of equal self-worth to show that women's perceived lower self-worth is not objectively accurate. To demonstrate this point, it is useful to consider the "persons cases" in common law. For sixty years in the late nineteenth and early twentieth centuries, women brought legal cases to claim rights on the grounds that statutes (for example, a statute that allowed someone to stand for election) referred to "persons," and that they, as women, should be included in the definition of "persons." After sixty years of refusing to view women as "persons," the common law courts were forced to relent. The newspapers, whose editors had previously belittled women's struggles, congratulated them on the progress they were making (Sachs and Hoff Wilson 1979, 40)! In other words, the judges' role in oppressing women was marginalized, and it appeared in the press as if women themselves had changed. Returning to Gauthier's objection to Hampton, recall that Gauthier asks, if we are not actually degraded, then how can we appear as such (2007, xiii)? This example illuminates Hampton's point that women were never less worthy of respect than men but were treated as if they were.[5]

A Feminist Analysis of Morality

The centrality in Hampton's work of the concept of the equal worth of persons forces us to focus upon the wrong of subordination and the ways in which it is inculcated. In this regard, I think Hampton's analysis is best understood as contributing to areas of feminist philosophy that analyze the psychological impact of our moral beliefs and that focus upon the details of injustice. I want to illustrate this by showing how her analysis intersects with other work in this area.

Miranda Fricker (2007) analyzes "epistemic injustice," which is made up of testimonial injustice and hermeneutic injustice. Testimonial injustice occurs when, for example, women's statements (in any setting, from intimate conversations to testimony in court) are not given the same credibility as men's—because women are viewed as either less knowledgeable or less truthful (or both). Hermeneutic injustice occurs when it is difficult to express a perception or experience within the prevailing language or cultural beliefs. For example, prior to the coining of the term "sexual harassment" and its use in law, it was difficult for women to explain this demeaning (and, in Hampton's terms, potentially "diminishing") experience. These two types of "epistemic," or epistemological, injustice are likely to reinforce each other, because someone who is stereotyped as either ignorant (as relying upon "women's intuition" rather than "facts," for example) or deceitful is even less likely to be believed if she has difficulty explaining herself and exhibits social cues of self-doubt. This may well affect her self-esteem, but, as Hampton makes clear, the real problem is the inculcation of the view that she can be treated as subordinate, thereby undermining the basis for her self-respect. The feeling of shame that accompanies the undermining of one's sense of self-worth also makes it less likely that such subordination will be challenged, especially if it is associated with experiences that cannot be shared easily with others.

Like Hampton, Fricker analyzes the psychological effects that result when such injustice follows someone throughout her life (because of her gender or race, for example). It means not only that she is denied credit for her insights but, worse, that she can start to lose faith in her judgment of events, which can then affect her moral judgments. This description of the psychological impact of being subjected to epistemological injustice is important because it illustrates how a particular type of injustice can become self-reinforcing. Fricker shows that there is a vicious circle in which a woman's lack of faith in her own judgment is then conveyed to others as she talks. She is less likely to be believed, thereby reinforcing her self-doubt. In this regard, Fricker's argument is similar to Hampton's descriptions of treatment (from serious crimes to petty daily exploitation) that has the capacity to make someone feel diminished. Both describe the subtleties of historically located injustice, which are self-reinforcing mechanisms that undermine self-esteem but also, more worryingly, self-respect and a sense of self-worth.[6]

The paucity of Rawls's psychological assumptions regarding our ability to be moral can be illustrated by comparison to Hampton's analysis. This is particularly relevant to his view of the family. As feminists, among them Elizabeth Brake in this volume, have made clear, children raised in a family that treats women as akin to servants are learning a lesson about treating others as subordinate. If the liberal state sanctions a particular view of the public-private divide in which injustice within the family is ignored by the state, then this injustice is perpetuated. The same applies to the perpetuation of subordination in the workplace.

Feminist Criticisms of Hampton's Contractarianism

I see Carole Pateman (1988, 1995, 2002) as the most important feminist critic of contractarianism in general. Pateman illustrates how, in modern societies, relations that were previously feudal (based upon status) came to be governed by contract. Employment contracts and marriage contracts (particularly between 1840 and 1970 in the West, in the latter case) thus became the way in which subordination was managed. The contract is therefore an odd concept to employ in discussion of free and equal persons, and whether it evokes the right imaginative insight in our culture is an empirical question. As Pateman rightly points out, however, Hampton's position does not rely upon contract but upon the concept of free and equal persons itself.[7] I thus agree on that point with Mills's (Pateman and Mills 2007, 21) argument that Pateman's position is compatible with that of Hampton, but I differ in what I draw from Hampton. I value her idea of keeping in play the idea of contract (or, better, of nonsubordination) in the consideration of whether actual relationships inculcate subordination.

Pateman (2002, 21) also complains that Rawlsian methodology reduces *political* struggles—struggles over who has a voice in everyday life—to discussions of abstract morality. While I share Pateman's concern, I think that Hampton traces the process whereby "moral" beliefs can affect the issue of whose voice is heard, particularly regarding the ethos of self-sacrifice in women's relationships that are based upon the norms of traditional marriage. A full understanding of the genealogy of such "moral" claims requires the sort of historical and political analysis that Pateman provides. What Hampton offers is a moral framework—based upon the axiom of equal self-worth—within which to critique such "moral" claims.

In addition, feminists have rightly criticized the appeal to abstract conceptions such as free and equal persons on the grounds that political and legal philosophers who employ this image have historically had men in mind as the norm. Women have been treated ambiguously, included either as part of "man" or as an inferior instantiation of the species. For example, in the contractarian tradition, Rousseau's *Emile* does not view women as able to be either natural men or citizens. The natural man should be able to stand up to public opinion, whereas women are to be defined by it and are to be trained to bear subordination. Similarly, Kant, while emphasizing independence of opinion in *What Is Enlightenment?* (1991a), is clear that men should speak for women in public (2006, 103).

In "Feminist Contractarianism," Hampton does not need to appeal to a contract or to some nonexistent gender-neutral view of "the person," nor does she rely fully upon Kantian metaphysics. The basis of her argument can be reduced to the axiom that it is morally wrong to treat some humans as if they were subordinate. The rest of her feminist arguments that I have discussed follow from that premise. I would maintain that "the person" (somehow both gender-neutral and also male) is only employed to focus upon everyday subordination. I refer to subordination rather than exploitation in order to hold on to the distinction for analysis, while recognizing their close association (Sample 2003). Subordination is the treatment of some persons as if they were of lesser worth than others. It is useful to consider the Marxist view of exploitation—as the extraction of surplus value—because it allows for greater historical analysis, as illustrated in the way that Pateman (2002) details the links between marriage and employment contracts, in which property in the person is exchanged.[8] However, an analysis of exploitation alone elides the problem of subordination that arises when the fiction of "property in the person" is accepted, thereby creating the roles of both employer and employee and of traditional husband and wife through contract. Pateman illustrates the fact that women are in a position to be exploited because they are viewed as subordinate when they exchange consortium for material support. However, as Hampton shows, exploitation can then reinforce subordination because it acts on a daily basis as a reminder of the claim that one is not worthy of equal respect, which may, over time, start to be (wrongly) believed (1988a, 39).

Conclusion

The ending of one of Hampton's papers contains an interesting twist. After she describes wrongdoing as involving defiance, a refusal to admit what is right, she suddenly expresses admiration for such a willingness to take on the world: "To appreciate how our capacity to defy authoritative norms may have helped us survive—as individuals, as a people, as a species—might enable us to become reconciled to something that exponents of various sorts of norms would argue that we ought bitterly to regret. And perhaps it is because we do see this capacity as important to our survival that it becomes hard to shake the thought that the first two human beings were impressive in the way they took on God" (2007b, 106–7). Hampton's contractarian test is set up to focus upon situations in which women rebel against both emotional ties and their perceived "duty" of self-sacrifice. In doing so, she keeps the concept of a contract in play—in contrast to Rawls's one-time thought experiment—to allow her to think about what it would be like for everyone to treat themselves, as well as others, as if they were free and equal. This allows her to focus upon the "primary good" of self-respect and to critique the ways it can be undermined.

The importance of self-respect is central to the work of both Rawls and Hampton, but Hampton provides us with a much richer analysis of its meaning. It is telling that Rawls initially fails to distinguish self-respect from self-esteem. This distinction is central to the contrast that Hampton draws between the morality of Kant and Hobbes within the contractarian tradition. For her, it is important to claim that women are equally worthy of respect even when systematically treated as subordinate, to distinguish equal worth from the fluctuating value attributed to us by others. The limits of Rawls's analysis of self-respect are reflected in his characterization of shame as associated with an (ahistorical) individual who either has a shoddy plan of life or an inability to carry out his or her life plan. Hampton's feminist philosophy is more nuanced in illustrating the relationship between psychological experiences of shame and their political context. She demonstrates how shame can be associated with self-identity—when women are treated in demeaning ways that reinforce gender stereotypes, for example. This opens up concerns about the subtle ways in which subordination is perpetuated at a mundane level within everyday relationships. Hampton's work therefore shifts the focus of justice to the continuing problem that injustice in the family poses for Rawls's thought.

Notes

1. For Hampton's argument that Rawls does not rely upon the contract but instead provides a method of practical reasoning that could be performed by one person, moving him closer to Kant, see Hampton 1980.

2. "For he that should be modest and tractable, and perform all he promises in such time and place where no man else should do so, should but make himself *a prey to others,* and procure his own certain ruin, contrary to the ground of all laws of nature which tend to nature's preservation" (Hobbes 1994, 99, emphasis added).

3. The Aristotelian principle states that "other things equal, human beings enjoy the exercise of their realized capacities (their innate or trained abilities), and this enjoyment increases the more the capacity is realized, or the greater its complexity" (*TJ,* 426).

4. For an analysis of this point in relation to law, see Richardson 2007, 42–44.

5. Given Hampton's debt to Kant, it is tempting to view her argument as an application of Kant's antinomy of freedom, i.e., that women are free in the ideal realm but not empirically. However, Kantian metaphysics is not necessary. It could simply be argued that the historical treatment of women was not inevitable. Although she would not have approved, Hampton could be viewed (against the grain of her realist view of morality) as contributing to a feminist genealogy of morality.

6. Fricker's concern dovetails with Bartky's (1990) analysis of how moral judgment can be compromised when one's beliefs conflict with those of the person for whom one is caring.

7. As indicated above, Hampton 1980 argues that Rawls does not rely upon contract itself. See also Hampton's critique of Rawls's overlapping consensus and her insistence that sex and race discrimination are in themselves wrong (1989, 813–14).

8. Pateman argues that the fiction that employees own their "property in the person," such as the ability to work that is then exchanged for a wage, allows them to appear to be equal in the polis and yet subordinate in the workplace. Employees are giving employers the right of command for a limited time, while appearing to exchange a commodity. Bartky 1990 explores this use of Marxist ideas to think about the phenomenology of both exploitation and subordination.

8

Liberal Feminism

Comprehensive and Political

Amy R. Baehr

As we have seen in the introduction to this volume, in PL Rawls argues for a turn in political philosophy from conceiving liberalism as a comprehensive moral doctrine to conceiving it as a public political philosophy. This reflects the conviction that coercive state action is justified—when constitutional essentials and basic justice are at stake—only if supported by "public reasons" (PL, 227–30). Public reasons are not the particular reasons of any one comprehensive moral doctrine. They are reasons sharable by the many reasonable comprehensive moral doctrines citizens hold. Some

---

This chapter is dedicated to my mother, Annie Baehr. For helpful discussion, thanks are due to Andrew Altman, Asha Bandary, Christie Hartley, Kevin Melchionne, Ira Singer, Kathleen Wallace, Lori Watson, Hofstra University's philosophy colloquium, and the Long Island Philosophical Society.

feminist liberals have recommended that feminists develop feminism as a public political philosophy (Baehr 2008; Brake 2004; Hartley and Watson 2010; Lloyd 1998, 209–210; McClain 2006; Nussbaum 2000a, 56; Nussbaum 2003, 511). Feminism as a public political philosophy—which I call "public political feminism"—is a set of feminist ends along with the public reasons that support them.[1]

What does the possibility of public political feminism mean for *liberal* feminism? Is liberal feminism to be identified with public political feminism? I suggest here that Rawls's comprehensive-political distinction makes possible a complex account of liberal feminism. According to this account, liberal feminism can be a comprehensive moral doctrine.[2] This is how it is commonly portrayed. But also, as I suggest above, liberal feminism can be a public political philosophy. If liberal feminism can be a public political philosophy, then one can count as a liberal feminist even if one rejects liberal feminism as a comprehensive doctrine. Say that you reject liberal feminism as a comprehensive doctrine because you endorse some other comprehensive feminist doctrine, for example, Jewish feminism, or ecofeminism. As long as your Jewish feminism or ecofeminism gives you reason to endorse public political feminism, you count as a kind of liberal feminist. It is an open question whether there are nonliberal feminist comprehensive doctrines that give adherents reason to endorse public political feminism (or whether there could be after a period of reflection and revision). I do not establish here that there are any. But if there are, then the liberal feminist tent is broadened.

This complex account of liberal feminism leads us to ask about the relationship between comprehensive feminist doctrines and feminism as a public political philosophy. While I conjecture that there are nonliberal feminist comprehensive doctrines that give adherents reason to endorse public political feminism, I argue in this chapter that there is at least one *liberal* feminist comprehensive doctrine that fails to provide such a reason. This seems counterintuitive and ought to be of interest to those of us who would like to hold liberal feminism both as a comprehensive doctrine and as a public political philosophy. If one is to hold them both, then there will have to be a comprehensive liberal feminism that has reasons for public political feminism.

In the first section below, I discuss the relationship between comprehensive doctrines and public political philosophy generally. I show that endorsement of public political feminism is not a rejection of the many feminist comprehensive doctrines, though it is a constraint on them. If one endorses public political feminism, then one may hold only a feminist comprehensive doctrine that has a reason for the limits of public reason. In the second section, I give a sketch of public political feminism. I focus on its ends and the public reasons for them. In the third section, I explore a kind of comprehensive liberal feminism that does not provide its adherents with reasons for endorsing public political feminism as the correct account of the coercive uses of state power to feminist ends. I also make the related point that this feminism is susceptible to three serious feminist criticisms. In the fourth section, I present a different comprehensive liberal feminism, one that has a reason for endorsing public political feminism and is responsive to the three serious feminist criticisms. I conclude with some remarks on the complex account of liberal feminism.

Comprehensive Moral Doctrines and Public Political Philosophy

What Rawls calls "the background culture" of society is "the culture of daily life"; it includes citizens' "comprehensive doctrines of all kinds—religious, philosophical, and moral" (PL, 14). Comprehensive doctrines are accounts of "what is of value in human life, ideals of personal character, as well as ideals of friendship and associational relationships and much else that is to inform conduct" (13).

Some comprehensive doctrines are feminist. Call to mind some examples of feminist ideals that inform personal and associational life: parenting guided by ideals of character such as androgyny, gender playfulness and nonconformity, or a character ideal found in an ethics of care; domestic and intimate life guided by ideals like gender egalitarianism, respect for sexual difference, or gender liberty; occupational choices guided by the value of women's caring work, the importance of women's independence and accomplishment, or solidarity with the world's women; spiritual life guided by a focus on the feminine qualities of the divine, or a determination to counteract the distortion caused by gendering God; charitable work, organizing, or electoral politics aimed at crafting caring communities, or ensuring that benefits and burdens in society are distributed in ways that are gender-just. These examples reveal a diversity of feminist comprehensive doctrines in the background culture of our society. We see this in the rich history of disagreement among feminists. There is no one comprehensive doctrine held by all feminists; nor is one feminist comprehensive doctrine a common denominator for the rest. And there is no social authority—as there is for, say, Catholicism—that claims to settle the disagreements. Complicating matters, feminisms rarely purport to tell the whole story about what is of value in human life, or the whole story about ideals of character and associational life. This is why we often find compound forms of feminism, like Jewish feminism and ecofeminism.

It should come as no surprise that there is no consensus among feminists concerning the correct comprehensive feminist doctrine. Following Rawls, we note that under conditions of freedom of thought, expression, and association, individuals come to diverse conclusions about questions of value (*PL*, 54–58). This is to be expected also within social movements (like the women's movement). Just as it is senseless for the state to impose one comprehensive doctrine on all citizens, it is senseless for participants in social movements to expect this kind of conformity. Thus the point is not that feminism can be understood as a *singular* comprehensive doctrine. It is that the many feminisms can be understood as *many* comprehensive doctrines.

Rawls's argument for political liberalism does not suggest that individuals must give up their comprehensive doctrines. He recommends that political philosophy conceive of individuals as capable of conceiving, revising, and being guided by a comprehensive conception of the good life (*PL*, 19). The point of public political philosophy is to allow for the exercise of this capacity, and to allow for the plurality of reasonable comprehensive doctrines to which it leads. Public political philosophy does this by showing that a just and stable political order need not be grounded in one particular doctrine to the exclusion of the many others, but may be grounded in shared public values.

While political liberalism does not necessarily require the abandonment of citizens' comprehensive moral doctrines, it does recommend a division of authority between public political philosophy and comprehensive doctrines. The former explains the ends to which coercive state power may be put, when constitutional essentials and basic justice are at stake, while comprehensive doctrines explain, as we have seen, "what is of value in human life, . . . ideals of personal character, . . . ideals of friendship and associational relationships and much else" (*PL*, 13). This way of describing comprehensive doctrines suggests that they are nonpolitical, inhabiting exclusively the "background culture" of society (14). This flies in the face of the obvious fact that many comprehensive doctrines—indeed, many feminist ones—are quite political, in the sense that they involve claims about the proper distribution of power in society generally, are articulated by citizens in the public realm of political debate and discussion, and include claims about how state power should be used. Rawls recognizes that many comprehensive doctrines are political. It is this political nature of many comprehensive doctrines that creates the problem that political liberalism is proposed to solve.

Indeed, Rawls explains that citizens' comprehensive doctrines have a legitimate role to play in the public sphere, understood as the sphere of discussion and debate about issues of common concern—including about constitutional essentials and basic justice (*PL*, 247–48). In a well-ordered society, contributions from comprehensive doctrines are excluded only from the public realm narrowly conceived. As Charles Larmore explains, reasons drawn exclusively from comprehensive doctrines are excluded, and "the ideal of public reason . . . ought to be understood as governing only the reasoning by which citizens—as voters, legislators, officials, or judges—take part in political decisions (about fundamentals) having the force of law" (Larmore 2003, 383). Thus in a well-ordered society, comprehensive doctrines can be said to belong to the wider public sphere of debate and discussion, but not to the narrower public sphere in which authoritative decisions are taken about coercive uses of state power.

Comprehensive doctrines play a somewhat different role in societies that are not well ordered. Some such societies lack, or largely lack, public reason. In such societies it can be important for citizens to reason publicly from their comprehensive doctrines, but to do so "for the sake of the ideal of public reason" (*PL*, 251), that is, for the sake of developing shared political values with which to collectively manage society's affairs concerning constitutional essentials and basic justice. Other not-well-ordered societies have public reason but lack substantial consensus on the meaning of their public values. There is reason to believe that Western democracies like the United States are not well ordered in this way. In such societies, reasoning publicly from comprehensive doctrines can be *for the sake of public reason*, to move a political community toward the "most reasonable" understanding of its political values (227).

Feminist scholars and activists have revealed many faces of gender injustice, and have shed light on the inability of dominant understandings of our public political values to illuminate that injustice. This has led some feminists to argue that feminist political thought is best understood as a rejection of those values. A women's movement guided predominantly by this self-understanding acts imprudently because it misses out on the opportunity to influence the interpretation of our public political values. But it also acts immorally because, even in a not-well-ordered society, to fail to reason at least for the sake of public reason is to show a kind of civic disrespect; it is to treat reasonable others as if they did not deserve to live under conditions they can affirm (*PL*, 217). Also, if feminists fail to participate in the development of public reason, we may well fall seriously short of the most reasonable understanding of our public values. To do this would be to deny citizens the justice they deserve.

Public Political Feminism

One endorses public political feminism if one is a feminist and believes that coercive state power, the power of "We the People," may not justly be used, even to feminist ends, unless it can be supported by public reasons.[3] (Because ours is not a well-ordered society, one endorses public political feminism if one believes that reasoning publicly from one's comprehensive feminist doctrine is permissible only *for the sake of public reason*.) I conjecture that there are many feminists, holding a variety of comprehensive moral doctrines, who believe that one should limit one's demands on state power in this way. On the complex account of liberal feminism, they count as liberal feminists even if they reject liberal feminism as a comprehensive doctrine.

The feminist ends to which state power may be put, according to public political feminism—that is, its content—cannot be stated completely and once and for all. It is a matter of what we can construct, and thus depends to a large degree on dedication and ingenuity. The literature[4] offers arguments for the following ends (note that these are some of the most important ends of feminism): Women should be free from coercion and violence, including and especially domestic and sexual violence, and violence enforcing domestic or sexual servitude. The state should actively protect gender liberty; fixed sex roles, sexual identity, and sexual orientation should not be enforced. The state should protect and promote the development of girls' and women's personal and political autonomy, and should reject relations of domination and subordination in the home. And the disadvantages that women suffer as a result of their disproportionate share of the burdens of reproduction must be remedied.[5]

In the literature we find two (compatible) ways to show that public reason supports ends like these.[6] One is to show that the best understanding of a particular public political value supports using state power to some feminist end. Consider some examples. The public political value of equality of opportunity requires rejecting workplace regulations that reflect an endorsement of a traditional division of labor, that is, that reflect a particular (patriarchal) comprehensive doctrine.[7] Its most reasonable interpretation suggests that the benefits and burdens of social cooperation include the burdens of reproduction; thus they too must be distributed fairly (Lloyd 1998, 218; McClain 2006, 92). The public political values of freedom of association and toleration require rejecting public policy that relies on heterosexist comprehensive doctrines (Brake 2004, 293); these values lead instead to support for same-sex marriage (McClain 2006, 156). Sex equality itself is a public political value that has broad implications for public policy; for example, it requires that the state not practice viewpoint neutrality "as between sex equality and its opposite; [the state] must put a thumb on the scales in favor of" sex equality (Case 2009, 397–98). This means state opposition to relations of subordination and domination in the family. This strategy of using public political values to support feminist ends is not new, of course. What is particular to public political feminism, however, is the claim that feminists have a moral obligation to appeal only to shared public values.

Another way to show that public reason supports feminist ends is to argue that the very activity of public reasoning presupposes the status of women as equal citizens, which in turn requires the realization of many of the ends listed above. Christie Hartley and Lori Watson argue that public reasoning presupposes civic respect among citizens that is incompatible with "pervasive social hierarchies" (Hartley and Watson 2010, 9). Thus state action to undermine those hierarchies is justified.

Of interest here is this question: What reason could a feminist have for limiting herself to public reasons, for recognizing the "duty of civility," the duty to explain to other citizens "how the principles and policies [she] advocate[s] . . . can be supported by the political values of public reason" (PL, 217)? In the next two sections I look at whether there is a comprehensive liberal feminist doctrine that gives adherents a reason to accept this duty.

Comprehensive Liberal Feminism I: Popular Liberal Feminism

There is more than one comprehensive liberal feminist doctrine. The comprehensive liberal feminism described here I call "popular liberal feminism" (PLF).[8] PLF has at its core an ideal of character, an ideal of domestic and intimate association, and a way of conceiving work. It includes a vision of solidarity among women and an agenda for the women's movement. As part of the latter, it includes an account of how power—including coercive state power—may and should be used to feminist ends.

Consider first the ideal of character. PLF recommends that a woman be independent and self-sufficient. This means being in the habit of distinguishing herself and her own interests from others'; developing and exercising her own talents; prioritizing her own aims; advocating for her fair share and expecting reciprocity from others; satisfying her needs and wants through her own endeavor; and insisting on her own separate value within contexts that transcend the family and intimate association. This ideal of character conflicts strongly with the character ideal of traditional femininity. That ideal recommends that women remain dependent on family, particularly on male members; define self in relation to others; take others' interests as their own; promote the development of others' talents at their own expense; and accept less than their fair share.

The popular liberal feminist character ideal of independence and self-sufficiency is coupled with an ideal of intimate and domestic association. According to this ideal, intimate and domestic association should be fair. One version of this ideal says that an intimate or domestic association is fair if the benefits that flow *from* each partner *to* the other are on par, that is, if each gives as much as she gets (Hampton 2007a; Radzik 2005; Sample 2002; Dimock 2008; I do not claim that these authors endorse PLF). On this view, affective benefits from a relationship that do not flow *from the other* but flow instead *from one's own caring nature*—are not counted (Radzik 2005, 51). This ideal requires distributing (nonaffective) benefits and burdens of a relationship exactly equally, so that domestic labor (a burden traditionally carried disproportionately by women) and income and leisure (benefits traditionally enjoyed disproportionately by men) are shared fifty-fifty. This ideal of domestic and intimate association rejects traditional domestic gender arrangements in which women carry more than their fair share of burdens and get less than their fair share of benefits—that is, in which women give to associated others more than they receive from those others.

As the emphasis on the division of labor in the home and access to wage work indicate, PLF is particularly concerned with the role that work plays in women's lives. It describes domestic work (housework and caregiving) as a burden of domestic association that, because it is unpaid, regularly renders women dependent upon a domestic partner who can generate income, and does not earn recognition in the wider community. Also, such work is regularly assigned to women and girls by social norms, becoming an obstacle to their pursuit of freely chosen ends. PLF describes wage work, by contrast, as a benefit to get one's fair share of, as an avenue to independence, self-sufficiency, and recognition in the wider world.

PLF includes an account of feminist solidarity: to be in solidarity with women is to want for them, and to help them live, lives characterized by independence and self-sufficiency, fairness in domestic and intimate association, and wage work outside the home. PLF holds that the women's movement's task is to operationalize this form of solidarity. Activists in the women's movement seek to realize their ends through the exercise of power. Informal power is exercised when activists seek to persuade—through education campaigns, demonstrations, and so on—but also when they create institutions that expand women's options, for example, day-care centers, rape crisis centers, microcredit for women's entrepreneurship, and women's professional associations. PLF holds that this power should be exercised so as to realize its ideals.

The women's movement also seeks to harness the coercive power of the state to its ends. It engages in electoral politics, lobbies elected representatives and regulatory agencies, and brings cases in the courts. PLF holds that coercive state power may and should be used to promote its ends. It endorses measures like these, which target *girls:* the legal prohibition of child marriage; access to contraception and abortion for girls without parental consent or notification; entitlement to an education that promotes independence and self-sufficiency and includes sex education, instruction in the legal equality of women, and equal access to sports. Legal measures targeting adult *women* include prohibiting discrimination against women, pregnant women, and women with dependency obligations at home, in hiring, pay, and seniority; comparable worth; affirmative action; flextime; time off for caregiving responsibilities; a right to express milk or breastfeed on the job; on-site day care; improved pay and benefits for part-time work; requirements that men take time off to care for newborns or newly adopted children; informing those to be married that current law rejects women's domestic servitude and condemns domestic violence; state-funded caregiver accounts; and giving non-wage-earning spouses a legal right to half of marital assets, including half of a spouse's income. PLF holds that measures to protect adult women are important for girls because a girl's chance to become independent and self-sufficient, to insist on her fair share, and to pursue work outside the home is reduced when the adult women in her family fail to model these virtues.

PLF gives adherents this reason for the legal measures it recommends: They are conducive to the particular, substantive way of life PLF recommends, namely, a life characterized by women's independence and self-sufficiency, fairness in intimate and domestic association, and work outside the home. This way of life, it holds, is the best life for women. Of course, this reason will not move someone whose comprehensive doctrine recommends a different way of life. A popular liberal feminist might have a pragmatic reason for occasionally finding public reasons—that is, reasons that those who hold different comprehensive doctrines can also endorse. Occasional coalitions with ideological opponents can be useful. But PLF attributes disagreement about the best life to nefarious causes—for example, to patriarchal ideology—and thus does not recognize what Rawls calls "the burdens of judgment" (PL, 56–57). To apply Rawls's words, PLF supposes that disagreement is not an understandable result of the free use of human reason but is due to "ignorance and perversity" and to "rivalries for power, status, or economic gain" (58). So if one wants to hold a comprehensive liberal feminist doctrine that gives reasons for public political feminism, PLF must be rejected.

There are additional reasons for feminists to reject PLF. First, note that all human communities are characterized by relations of dependency. Not only is each of us dependent on care provided by others when we are children (and many are dependent when temporarily or permanently disabled), but those who care for dependents rely on still others for their own support. Eva Kittay calls this the system of "nested dependencies" (1999, 66–68). Feminist theorists have described in great detail the kinds of work, the traits of character, and the forms of association involved in systems of nested dependency (Ruddick 1989; Held 1987; Tronto 1993). Such feminists acknowledge that character traits of traditional caregivers, and traditional forms of domestic association involved in caregiving, have been sources of women's oppression, but they still urge us to acknowledge the intrinsic value of being cared for, as well as the intrinsic value of the traits of character that make good caregiving possible. They wisely recommend the reinvention of intimate and domestic association, and the reevaluation of work, in ways that support both caregiving and caregivers. But PLF perpetuates the devaluing of caregiving work and the character traits associated with good caregiving. This devaluing is a significant cause of women's disadvantage; and it diverts attention from seeing to it that those who need care are cared for, and that those who provide care are supported. Also, the devaluing of caregiving work leads PLF (implausibly) to exclude the possibility that a woman's life can be good precisely because of—not despite—the role that caregiving plays in it.

Second, while many privileged women's lives have been enhanced by recently won access to professional work, many other women have always had to work outside the home. For many less privileged women, work outside the home was and continues to be physically exhausting, harmful, boring, poorly paid, socially stigmatized, and involving little control over working conditions. As bell hooks writes, "Work has not been a liberating force for masses of American women" (1981, 146). As men's wages have remained stagnant or fallen, more and more women in heterosexual households have no choice but to do such work outside the home. The availability of low-wage women workers is a boon to more privileged women. When low-wage women workers take care of privileged women's dependents and their homes, privileged women are freed up to pursue enriching work outside the home. This also frees privileged women to volunteer and otherwise nurture the larger associations on which their communities depend. Low-wage women workers can rarely pass their domestic work on to someone else, and they struggle to nurture the communities in which they are embedded (Romero 1997). This suggests that PLF's emphasis on wage work outside the home supports the advantage of some women at the expense of other women—and this is dramatically so when we consider the global dimensions of women's wage work. On this view, PLF makes feminism at best irrelevant to many less privileged women, and at worst complicit in their exploitation (Holmstrom 2011; Fraser 2011; Eisenstein 2009; Romero 1997). So much for solidarity.

Third, as much recent feminist theory has emphasized (Shachar 2007), women and men alike are embedded in cultural (and many of us in religious) traditions that contribute significantly to our identities and to our thinking about what gives our lives value and how we ought to live. Our traditions are not monolithic. Nor are we their victims. We construct our lives within them, drawing on them, revising them, rarely rejecting them entirely. They are as constitutive as they are constraining. They are, as Rawls explains, exercises in practical reason. A comprehensive liberal feminism that asserts one supposedly best way of life for women against this existing diversity is at best irrelevant to many women. At worst, it is a threat to many women's identities and values (Shachar 2009, 152; see also Wolf-Devine 2004; Macklem 2003). The point of this feminist criticism of PLF is not to assert a noncritical cultural relativism. It is rather to warn against a kind of feminist hubris.

The comprehensive liberal feminist doctrine presented in the next section deals more carefully with these three feminist concerns: the fact of human dependency, the role of work in a good life, and the fact of cultural diversity. It also gives its adherents reason to endorse public political feminism as the correct account of the just uses of state power to feminist ends.

Comprehensive Liberal Feminism II: The Autonomy Account

Susan Okin writes, "Liberalism's central aim should be to ensure that every human being has a reasonably equal chance of living a good life according to his or her unfolding views about what such a life consists in" (1999, 119).⁹ A comprehensive liberal feminist doctrine with this aim at its heart is concerned not with promoting a particular, substantive way of life for women but with making it possible for women to live lives that are good by their own lights. It holds that being able to live a life that is good by one's own lights, having personal autonomy, is a minimal condition of a good life. And it is that condition with which the women's movement should concern itself.¹⁰

To value women's living lives that are good by their own lights is to value what facilitates this. What facilitates it is valued as a means to an end. Though ends may be the same, not all contexts are the same, so the means will differ. Means must be tailored to particular contexts. We may not presume that there are any particular means that facilitate equally each woman's being able to live a life that is good by her own lights. So when we inquire into conditions or measures that we think might facilitate women's living such lives, we must ask questions like these: Which women's personal autonomy is this likely to facilitate? Under what socioeconomic or cultural conditions? By what mechanisms? What consequences are likely to result, apart from freeing some women to live lives that they value? Who might choose these consequences over the status quo ante, and who might reject them? Might this increase personal autonomy in one way, or for one kind of woman, even as it reduces personal autonomy in another way, or for another kind of woman?

According to this comprehensive liberal feminist doctrine, which I call the autonomy account (AA), feminist theory and the women's movement are forums for the discussion of questions like these, for the discovery of the conditions under which diverse women in diverse contexts can exercise personal autonomy. Discussion in these forums, which aims at discovering these conditions, must include all women if it is to be a *women's* movement. But it need not be one conversation. Results are likely to be better if there are multiple conversations. Also, discussion does not aim at consensus about what measures facilitate equally *all* women's autonomy.[11] Such an aim assumes (which it should not) that there are means that facilitate all women's autonomy equally. Nor should discussion aim to secure women's autonomy once and for all. Conditions change. Discussion aims at discovering the diverse and possibly incommensurate conditions necessary for diverse women in diverse and changing contexts to live lives that they value.

Some feminisms mistake conditions and measures that facilitate *some* women's personal autonomy for conditions and measures that facilitate *all* women's personal autonomy. PLF makes this mistake. Its devaluing of the work of care and the identities of caregivers, its ignorance of the great costs that actual wage work imposes on many women and their communities, and its insistence on independence and self-sufficiency as the sine qua non of the good life reveal this. To be sure, emphasizing the importance of women's work outside the home, insisting on fifty-fifty sharing of domestic burdens, and many other measures that PLF recommends have played and continue to play an important role in many women's being able to live lives that they value. But AA urges us to recognize the contextual nature of the value of the way of life PLF recommends. AA urges us to focus on diverse lives, especially on the lives of women for whom popular feminist ideals are not necessary conditions for a life that is good by their own lights, but may even be in conflict with it.

On this account, to be in solidarity with women is not to urge them to live a particular way of life but to empower them to develop their "unfolding views" about what a good life consists in, and to live something like the life they value. This is a solidarity of empowerment, support, and respect, a vision of solidarity among women that can guide a women's movement that is economically, culturally, geographically, and generationally[12] heterogeneous (see Meyers 2004, 205).

While we must not elevate conditions that facilitate some women's autonomy as if they facilitated all women's, we can venture a few general necessary conditions. Women must be free from violence and the threat of violence (Cudd 2006, 85–118; Brison 1997); free from social practices and laws that coercively steer women into socially preferred ways of life (Okin 1989a, 170ff.; Alstott 2004; Meyers 2004; Cornell 1998, x); free from material deprivation (Cudd 2006, 119–54) and imagination-stifling cultural homogeneity, so that they have options (Cudd 2006, 234; Alstott 2004, 52); and they must have certain capacities—for example, the ability to assess their preferences and imagine life otherwise (Meyers 2002, 168; Cudd 2006, 234–35; MacKenzie 1999).

This feminism does endorse uses of state power to feminist ends. It recognizes that society's structure is, to a significant degree, determined by state action, and it acknowledges that it is unlikely that this structure will facilitate women's autonomy if women do not participate in formal politics and use the tool of legal regulation. So, while it supports many of the legal measures advocated by PLF, it does so not because they are conducive to a particular feminist way of life but because they facilitate women's living lives that are good by their own lights. This leads AA, however, to reject some legal measures, namely, those grounded in a particular ideal of intimate and domestic association (for example, legally requiring that men take time off to care for newborns or newly adopted children, and giving non-wage-earning spouses a legal right to half of marital assets, including half of the wage-earning spouse's income) (Alstott 2004, 113; Wolf-Devine 2004). Also, because of its focus on the diversity of women and their contexts, this feminism has a healthy skepticism about legal measures. Legal measures tend to be one size fits all, and hard to change. This feminism wants to be more nimble than legal measure sometimes allow.

AA's healthy skepticism about state power leads in two directions. First, it emphasizes the importance of robust and inclusive public discussion, so that the legal regulation of the basic structure remains responsive to the needs of women. Second, AA focuses particular attention on the many and diverse activities through which women conceive and revise their conceptions of the good life, drawing on or remaining largely faithful to inherited cultural or religious traditions or moving beyond them, developing new "alternative emancipatory imagery" (Meyers 2002, 168) and fashioning new ways of being a woman and new kinds of relationships through experiments in living (Cudd 2006, 234).

Does this feminism give adherents a reason for accepting the duty of civility, the duty to limit calls for state power, even to feminist ends, to those that can be supported with public reasons?[13] This feminism gives its adherents a pragmatic reason. It says that when feminists are in the minority, public reason renders others unable to impose their nonfeminist way of life. But AA has a moral reason for the limits of public reason; it explains why feminists should limit themselves even if they are in the majority and have sufficient power to impose a way of life coercively. AA holds that a good life, by its very nature, cannot be imposed. This is because the minimal condition for a life's being good is that it is recognized as such by the person who lives it. This includes, of course, the fundamental conditions under which one lives—the basic structure of society. That too must be acceptable. Since AA holds this, it holds that feminists must offer public reasons for feminist uses of state power.[14] Thus AA gives adherents reason for the duty of civility.

Critics worry that, even enjoying the conditions of autonomy described above, women may continue to choose disadvantaging and oppressive arrangements, and that a women's movement limited to the value of autonomy would have little normative ground for criticism (Yuracko 2003). Some point to the phenomenon of deformed preferences: when attractive options are limited or arrangements unfair, people may develop preferences for those limits or for less than their fair share (Nussbaum 1999, 33; Cudd 2006, 180–83). It is clear that PLF has an advantage over AA here. PLF can simply say that it is wrong, for example, for arrangements to diverge from a fifty-fifty division of nonaffective benefits and burdens, or for women to prioritize the satisfaction of others' needs. AA cannot do this. The conditions AA insists on (the conditions for autonomy) do not rule out the possibility that a woman could choose, for example, to undergo a clitorectomy (Meyers 2004, 213), or to become a pornographic model (Cudd 2004, 58), a nun, or a submissive wife and mother.

There are two responses to this worry. (Neither will quell it entirely.) The first response is that AA is presented not as a complete account of the good life for women but as a comprehensive doctrine to guide a women's movement that is economically, culturally, geographically, and generationally heterogeneous. So AA is presented here as a political ethic for the women's movement, not as a full account of what makes a woman's life good.[15] The second response is to stress that diverse autonomous women will construct lives dedicated to a wide variety of goods. Feminist theorists and activists (like everyone else) have a limited moral imagination; so we cannot expect to be able to anticipate the nature of these lives, or even to fully appreciate them (Meyers 2004, 213). But our solidarity ought not to end with our ability to appreciate fully the goods to which others dedicate themselves.

Conclusion

If one wants to endorse liberal feminism both as a comprehensive doctrine and as a public political philosophy, one must be careful about which comprehensive liberal feminist doctrine one endorses. The autonomy account, unlike popular liberal feminism, gives adherents reason to endorse public political feminism as the correct account of the coercive uses of state power to feminist ends. AA is also more attractive than PLF for other feminist reasons, as we have seen. Its solidarity of empowerment, support, and respect can guide a women's movement that is economically, culturally, geographically, and generationally heterogeneous.

But one need not endorse a comprehensive liberal feminist doctrine in order to count as a liberal feminist. On the complex account of liberal feminism, one counts as a liberal feminist as long as one's comprehensive doctrine provides reasons for the duty of civility, and thus for public political feminism. Are there nonliberal feminist doctrines that provide such reasons? I conjecture that there are, or could be after a period of reflection and revision. But the complex account of liberal feminism is an invitation to those who hold the many feminist comprehensive doctrines to explore their values—the way I have explored comprehensive liberal feminism here—with an eye to whether they have reason, or after a period of reflection and revision could find reason, to participate in the development of our society's public reason. A women's movement invigorated by this call acts prudently, because it takes advantage of the opportunity to influence the interpretation of our public political values. But it also does the work of justice, as it may help us to come closer to the most reasonable understanding of our public values, and thus to give citizens the justice they deserve.

Notes

1. Its role is to guide efforts to use state power to feminist ends. Comprehensive feminist doctrines may guide other feminist activities.

2. As indicated below, there is more than one comprehensive liberal feminist doctrine.

3. That is, when constitutional essentials and basic justice are at stake, and when the reasoning leads directly to decisions' having the force of law.

4. See, for example, Baehr 2008; Hartley and Watson 2010; Lloyd 1998; McClain 2006; Nussbaum 2000a. For related work, see also Brake 2004 (but see 300); Case 2009; Laden 2001.

5. Is legalization of prostitution part of public political feminism? Baehr 2008 and Nussbaum 1999, 276–98, suggest that it is. For public reasons why it is not, see Hartley and Watson 2010, 16. What about violent pornography? Anthony Laden argues that calls for restriction are not unreasonable (2003, 148–49); Susan Brison argues that the best arguments for freedom of expression fail to show that violent pornography should not be limited (1998); Christina Spaulding argues that violent pornography can undermine the status of women as equal citizens (1988–89). For an excellent discussion of what a public reason argument for banning violent pornography needs to look like, see Eaton 2007.

6. Even when a particular feminist does not explicitly identify herself as contributing to public political feminism, her arguments may involve the offering of public reasons.

7. Brake argues that the legal regulation of the workplace "was historically formed on the basis of a gendered division of labor . . . [and] unduly promote[s] one conception of the good—that in which primary care-givers for young children stay home, while their partners work outside the home to support the family" (2004, 308).

8. I offer PLF, and a contrasting comprehensive liberal feminism in the next section of this chapter, to fix ideas. I do not claim that anyone holds these doctrines precisely as described. Nancy Rosenblum presents a related contrast in an intriguing paper on Susan Okin (Rosenblum 2009; on Rosenblum, see Baehr 2010).

9. I put aside what it means to have "a reasonably equal chance"; for more on equality in Okin, see Miller 2009. I do not claim that Okin embraced the feminism I develop in this section. For a reading of Okin that inspired this section, see Rosenblum 2009. The comprehensive feminism described in this section has strong affinities also with the work of Diana Meyers (2002, 2004) and Drucilla Cornell (1998).

10. This feminism is offered not as a complete account of a good life but as an account sufficient to orient the women's movement.

11. Benhabib 1992, 52, stresses the importance of not aiming at consensus.

12. Self-described "third wave" feminists have stressed that one generation's struggle is not necessarily another's (Baumgardner and Richards 2000).

13. Thanks to Andrew Altman for helpful discussion of this issue.

14. Unreasonable others may reject these reasons. But that need not concern us here.

15. Thanks to Ira Singer for discussion of this issue.

References

Abbey, Ruth. 2007. "Back Toward a Comprehensive Liberalism? Justice as Fairness, Gender, and Families." *Political Theory* 35 (1): 1–24.
———. 2011. *The Return of Feminist Liberalism*. Montreal: McGill–Queen's University Press.
Ahmed, Leila. 1993. *Women and Gender in Islam: Historical Roots of a Modern Debate*. New Haven: Yale University Press.
Alstott, Linda. 2004. *No Exit: What Parents Owe Their Children and What Society Owes Parents*. New York: Oxford University Press.
Americans with Disabilities Act of 1990. 42 U.S.C. http://www.ada.gov/statute.html.
Anderson, Elizabeth S. 1999. "What Is the Point of Equality?" *Ethics* 109 (2): 287–337.
Arneson, Richard. 1990. "Primary Goods Reconsidered." *Nous* 24 (3): 429–54.
———. 2000. "Disability, Discrimination, and Priority." In *Americans with Disabilities: Exploring Implications of the Law for Individuals and Institutions*, edited by Leslie Pickering Francis and Anita Silvers, 18–33. New York: Routledge.
Audard, Catherine. 2007. *Philosophy Now: John Rawls*. Montreal: McGill–Queen's University Press.
Baehr, Amy. 1996. "Toward a New Feminist Liberalism: Okin, Rawls, and Habermas." *Hypatia* 11 (1): 49–66.
———. 2007. "Liberal Feminism." *Stanford Encyclopedia of Philosophy*. Available online at http://plato.stanford.edu/entries/feminism-liberal/.
———. 2008. "Perfectionism, Feminism, and Public Reason." *Law and Philosophy* 27:193–222.
———. 2010. Review of *Toward a Humanist Justice: The Political Philosophy of Susan Moller Okin*, edited by Debra Satz and Rob Reich. *Social Theory and Practice* 36:525–33.
Barry, Brian. 2000. *Culture and Equality*. Oxford: Basil Blackwell.
Bartky, Sandra Lee. 1990. *Femininity and Domination: Studies in the Phenomenology of Oppression*. London: Routledge.
Baumgardner, Jennifer, and Amy Richards. 2000. *Manifesta: Young Women, Feminism, and the Future*. New York: Farrar, Straus and Giroux.
Benhabib, Seyla. 1992. *Situating the Self*. New York: Routledge.
Benn, Stanley. 1988. *A Theory of Freedom*. Cambridge: Cambridge University Press.
Berlin, Isaiah. 1971. "Two Concepts of Liberty." In Berlin, *Four Essays on Liberty*, 118–72. New York: Oxford University Press.

Bérubé, Michael. 1992. *Marginal Forces / Cultural Centers: Tolson, Pynchon, and the Politics of the Canon.* Ithaca: Cornell University Press.

Bojer, Hilda. 2000. "Children and Theories of Social Justice." *Feminist Economics* 6 (2): 23–39.

Bos, Henny M. W., and Frank van Balen. 2008. "Children in Planned Lesbian Families: Stigmatisation, Psychological Adjustment, and Protective Factors." *Culture, Health, and Sexuality* 10 (3): 221–36.

Brake, Elizabeth. 2004. "Rawls and Feminism: What Should Feminists Make of Liberal Neutrality?" *Journal of Moral Philosophy* 1 (3): 293–309.

———. 2010. "Minimal Marriage: What Political Liberalism Implies for Marriage Law." *Ethics* 120:302–37.

Brennan, Samantha, and Robert Noggle. 2000. "Rawls's Neglected Childhood: Reflections on the Original Position, Stability, and the Child's Sense of Justice." In *The Idea of Political Liberalism: Essays on Rawls,* edited by C. Wolf and V. Davion, 46–72. Lanham, Md.: Rowman and Littlefield.

Brettschneider, Corey. 2007. "The Politics of the Personal: A Liberal Approach." *American Political Science Review* 101 (1): 19–31.

Brickman, P., D. Coates, and R. Janoff-Bulman. 1978. "Lottery Winners and Accident Victims: Is Happiness Relative?" *Journal of Personality and Social Psychology* 36:917–27.

Brighouse, Harry, and Adam Swift. 2006. "Parents' Rights and the Value of the Family." *Ethics* 117 (1): 80–108.

Brison, Susan. 1997. "Outliving Oneself: Trauma, Memory, and Personal Identity." In *Feminists Rethink the Self,* edited by Diana Meyers, 12–39. Boulder: Westview Press.

———. 1998. "The Autonomy Defense of Freedom of Speech." *Ethics* 108:312–39.

Calhoun, Cheshire. 2003. *Feminism, the Family, and the Politics of the Closet: Lesbian and Gay Displacement.* Oxford: Oxford University Press.

Case, Mary Anne. 2009. "Feminist Fundamentalism on the Frontier Between Government and Family Responsibility for Children." *Utah Law Review* 2:381–406.

Chambers, Clare. 2008. *Sex, Culture, and Justice: The Limits of Choice.* University Park: Pennsylvania State University Press.

———. 2009. "Each Outcome Is Another Opportunity: Problems with the Moment of Equal Opportunity." *Politics, Philosophy, and Economics* 8 (4): 374–400.

Chambers, Clare, and Phil Parvin. 2010. "Coercive Redistribution and Public Agreement: Re-evaluating the Libertarian Challenge of Charity." *Critical Review of International Social and Political Philosophy* 13 (1): 93–114.

Chambers, Simone. 2010. "Theories of Political Justification." *Philosophy Compass* 5:893–903.

Clay, Daniel, Vivian L. Vignoles, and Helga Dittmar. 2005. "Body Image and Self-Esteem Among Adolescent Girls: Testing the Influence of Sociocultural Factors." *Journal of Research on Adolescence* 15 (4): 451–77.

Clayton, Matthew. 2006. *Justice and Legitimacy in Upbringing.* Oxford: Oxford University Press.

Cohen, G. A. 2008. *Rescuing Justice and Equality.* Cambridge: Harvard University Press.

Cornell, Drucilla. 1995. *The Imaginary Domain: Abortion, Pornography, and Sexual Harassment.* New York: Routledge.

———. 1998. *At the Heart of Freedom: Feminism, Sex, and Equality.* Princeton: Princeton University Press.

Cudd, Ann. 2004. "The Paradox of Liberal Feminism: Preference, Rationality, and Oppression." In *Varieties of Feminist Liberalism*, edited by Amy R. Baehr, 37–62. Lanham, Md.: Rowman and Littlefield.
———. 2006. *Analyzing Oppression*. New York: Oxford University Press.
Daniels, Norman. 1985. *Just Health Care*. Cambridge: Cambridge University Press.
———. 1990. "Equality of What: Welfare, Resources, or Capabilities?" *Philosophy and Phenomenological Research* 50 (supplement): 273–96.
Darwall, Stephen. 1977. "Two Kinds of Respect." *Ethics* 88:36–49.
Deigh, John. 1999. "Shame and Self-Esteem: A Critique." In *Moral Psychology and Community*, vol. 4 of *The Philosophy of Rawls: A Collection of Essays*, edited by Henry S. Richardson and Paul J. Weithman, 1–21. London: Garland.
Dimock, Susan. 2008. "Why All Feminists Should Be Contractarians." *Dialogue* 47:273–90.
Doppelt, Gerald. 2009. "The Place of Self-Respect in a Theory of Justice." *Inquiry* 52 (2): 127–54.
Eaton, A. W. 2007. "A Sensible Anti-Porn Feminism." *Ethics* 117:674–715.
Eisenstein, Hester. 2009. *Feminism Seduced: How Global Elites Use Women's Labor and Ideas to Exploit the World*. Boulder: Paradigm Press.
English, Jane. 1977. "Justice Between Generations." *Philosophical Studies* 31 (2): 91–104.
Exdell, J. 1994. "Feminism, Fundamentalism, and Liberal Legitimacy." *Canadian Journal of Philosophy* 24 (3): 441–63.
Eyal, Nir. 2005. "'Perhaps the Most Important Primary Good': Self-Respect and Rawls' Principles of Justice." *Politics, Philosophy, and Economics* 4 (2): 195–219.
Feinberg, Joel. 2000. "Disability and Illness." In *Americans with Disabilities: Exploring Implications of the Law for Individuals and Institutions*, edited by Leslie Pickering Francis and Anita Silvers, 244–54. New York: Routledge.
Flathman, Richard. 1987. *The Philosophy and Politics of Freedom*. Chicago: University of Chicago Press.
Fraser, Nancy. 2011. "Marketization, Social Protection, Emancipation: Toward a Neo-Polanyian Conception of Capitalist Crisis." In *Business as Usual: The Roots of the Global Financial Meltdown*, edited by Craig Calhoun and Georgi Derlugian, 137–58. New York: NYU Press.
Freeman, Samuel. 2007. *Rawls*. London: Routledge.
Fricker, Miranda. 2007. *Epistemic Injustice: Power and the Ethics of Knowing*. Oxford: Oxford University Press.
Gatens, Moira. 2004. "Privacy and the Body: The Privacy of Affect." In *Privacies: Philosophical Evaluations*, edited by Beate Rossler, 113–32. Stanford: Stanford University Press.
Gauthier, David P. 1987. *Morals by Agreement*. Oxford: Clarendon Press.
———. 2007. Foreword to *The Intrinsic Worth of Persons: Contractarianism in Moral and Political Philosophy*, by Jean Hampton, edited by David Farnham, ix–xiii. Cambridge: Cambridge University Press.
Giddens, Anthony. 1991. *Modernity and Self-Identity*. Cambridge: Polity Press.
Gilligan, Carol. 1982. *In a Different Voice: Psychological Theory and Women's Development*. Cambridge: Harvard University Press.
Grandin, Temple. 1995. *Thinking in Pictures: And Other Reports from My Life with Autism*. New York: Doubleday.

Green, Karen. 1986. "Rawls, Women, and the Priority of Liberty." *Australasian Journal of Philosophy* 64 (supplement): 26–36.
———. 2006. "Parity and Procedural Justice." *Essays in Philosophy: A Biannual Journal* 7 (1). Available at http://commons.pacificu.edu/eip/v017/iss1/4.
Hampton, Jean. 1980. "Contracts and Choices: Does Rawls Have a Social Contract Theory?" *Journal of Philosophy* 77 (6): 315–38.
———. 1988a. "Forgiveness, Resentment, and Hatred." In *Forgiveness and Mercy*, edited by Jeffrey G. Murphy and Jean Hampton, 35–87. Cambridge: Cambridge University Press.
———. 1988b. "The Retributive Idea." In *Forgiveness and Mercy*, 111–61. Cambridge: Cambridge University Press.
———. 1989. "Should Political Philosophy Be Done Without Metaphysics?" *Ethics* 99 (4): 791–814.
———. 1993. "Selflessness and the Loss of Self." *Social Philosophy and Policy* 10 (1): 135–65.
———. 1997. "The Wisdom of the Egoist: The Moral and Political Implications of Valuing the Self." *Social Philosophy and Policy* 14:21–51.
———. 1999. "Interview." In *Key Philosophers in Conversation*, edited by Andrew Pyle, 231–38. London: Routledge.
———. 2007a. "Feminist Contractarianism." In Hampton, *The Intrinsic Worth of Persons: Contractarianism in Moral and Political Philosophy*, edited by David Farnham, 1–38. Cambridge: Cambridge University Press.
———. 2007b. "Mens Rea." In Hampton, *The Intrinsic Worth of Persons*, 72–107. Cambridge: Cambridge University Press.
———. 2007c. "Righting Wrongs: The Goal of Retribution." In Hampton, *The Intrinsic Worth of Persons*, 108–50. Cambridge: Cambridge University Press.
Handley, Peter. 2003. "Theorising Disability: Beyond 'Common Sense.'" *Politics* (23) 2: 109–18.
Hartley, Christie, and Lori Watson. 2010. "Is a Feminist Political Liberalism Possible?" *Journal of Ethics and Social Philosophy* 5 (1): 1–21.
Haslanger, Sally. 1993. "On Being Objective and Being Objectified." In *A Mind of One's Own: Feminist Essays on Reason and Objectivity*, edited by Louise Antony and Charlotte Witt, 85–125. Boulder: Westview Press.
Held, Virginia. 1987. "Non-contractual Society: A Feminist View." In *Science, Morality, and Feminist Theory*, edited by Marsha P. Hanen and Kai Nielsen, 111–35. Calgary: University of Calgary Press.
Hines, Denise A., and Kathleen Malley-Morrison. 2005. *Family Violence in the United States: Defining, Understanding, and Combating Abuse*. Thousand Oaks, Calif.: Sage.
Hirschmann, Nancy J. 1992. *Rethinking Obligation: A Feminist Method for Political Theory*. Ithaca: Cornell University Press.
———. 2003. *The Subject of Liberty: Toward a Feminist Theory of Freedom*. Princeton: Princeton University Press.
———. 2008. *Gender, Class, and Freedom in Modern Political Theory*. Princeton: Princeton University Press.
———. 2013. "Freedom and (Dis)Ability in Early Modern Political Thought." In *Disabling the Renaissance: Recovering Early Modern Disability*, edited by Allison Hobgood and David Wood, 167–86. Columbus: Ohio State University Press.

Hirshman, Linda R. 1994. "Is the Original Position Inherently Male-Superior?" *Columbia Law Review* 94 (6): 1860–81.
Hobbes, Thomas. 1994. *Leviathan: With Selected Variants from the Latin Edition of 1668*. Edited by Edwin Curley. Indianapolis: Hackett.
Holmstrom, Nancy. 2011. "Against Capitalism as Theory and as Reality." In Ann E. Cudd and Nancy Holmstrom, *Capitalism, For and Against: A Feminist Debate*, 133–259. Cambridge: Cambridge University Press.
hooks, bell. 1981. *Ain't I a Woman: Black Women and Feminism*. Cambridge, Mass.: South End Press.
Kant, Immanuel. 1991a. "An Answer to the Question: 'What Is Enlightenment?'" In *Kant: Political Writings*, edited by Hans S. Reiss, 54–60. Cambridge: Cambridge University Press.
———. 1991b. *Observations on the Feeling of the Beautiful and the Sublime*. Translated by John T. Goldthwait. Berkeley: University of California Press.
———. 2006. *Anthropology from a Pragmatic Point of View*. Edited by Manfred Kuehn and Robert Louden. Cambridge: Cambridge University Press.
Kearns, Deborah. 1983. "A Theory of Justice—and Love: Rawls on the Family." *Politics: Australian Journal of Political Science* 18 (2): 36–42.
Kittay, Eva Feder. 1999. *Love's Labor: Essays on Women, Equality, and Dependency*. New York: Routledge.
Kohlberg, Lawrence. 1981. "Justice as Reversibility." In *Philosophy, Politics, and Society*, edited by Peter Laslett and James Fishkin, 257–72. Oxford: Basil Blackwell.
Korsgaard, Christine. 1996. *The Sources of Normativity*. Cambridge: Cambridge University Press.
Kristjansson, Kristjan. 1992. "What Is Wrong with Positive Liberty?" *Social Theory and Practice* 18 (3): 289–310.
Laden, Anthony Simon. 2000. "Outline of a Theory of Reasonable Deliberation." *Canadian Journal of Philosophy* 30:551–80.
———. 2001. *Reasonably Radical: Deliberative Liberalism and the Politics of Identity*. Ithaca: Cornell University Press.
———. 2003. "Radical Liberals, Reasonable Feminists: Reason, Power, and Objectivity in MacKinnon and Rawls." *Journal of Political Philosophy* 11 (2): 133–52.
———. 2010. "The Justice of Justification." In *Habermas and Rawls: Disputing the Political*, edited by James Gordon Finlayson and Fabian Frayenhagen, 135–52. New York: Routledge.
LaFollette, Hugh. 1980. "Licensing Parents." *Philosophy and Public Affairs* 9 (2): 182–97.
Larmore, Charles. 2003. "Public Reason." In *The Cambridge Companion to Rawls*, edited by Samuel Freeman, 368–93. Cambridge: Cambridge University Press.
Levey, Ann. 2005. "Liberalism, Adaptive Preferences, and Gender Equality." *Hypatia* 20 (4): 127–43.
Lewis, Jamie, and Lisa Pearce. 2008. "Adolescent Self-Image: The Role of Religion." Paper presented at the annual meeting of the American Sociological Association, Boston, Mass., 3 August.
Leys, Ruth. 2007. *From Guilt to Shame: Auschwitz and After*. Princeton: Princeton University Press.
Lloyd, Sharon A. 1994. "Family Justice and Social Justice." *Pacific Philosophical Quarterly* 75 (3–4): 353–71.

———. 1995. "Situating a Feminist Criticism of John Rawls's Political Liberalism." *Loyola of Los Angeles Law Review* 28:1319–44.

———. 1998. "Toward a Liberal Theory of Sexual Equality." *Journal of Contemporary Legal Issues* 9:203–24.

MacKenzie, Catriona. 1999. "Imagining Oneself Otherwise." In *Relational Autonomy: Feminist Perspectives on Autonomy, Agency, and the Social Self*, edited by Catriona MacKenzie and Natalie Stoljar, 124–50. New York: Oxford University Press.

MacKinnon, Catharine. 1987. *Feminism Unmodified*. Cambridge: Harvard University Press.

———. 1989. *Toward a Feminist Theory of the State*. Cambridge: Harvard University Press.

———. 2005. *Women's Lives, Men's Laws*. Cambridge: Harvard University Press.

———. 2006. *Are Women Human? And Other International Dialogues*. Cambridge: Harvard University Press.

Macklem, Timothy. 2003. *Beyond Comparison: Sex and Discrimination*. Cambridge: Cambridge University Press.

Mallon, Ron. 1999. "Political Liberalism, Cultural Membership, and the Family." *Social Theory and Practice* 25 (2): 271–97.

McClain, Linda C. 1991–92. "'Atomistic Man' Revisited: Liberalism, Connection, and Feminist Jurisprudence." *Southern California Law Review* 65:1173–1264.

———. 2006. *The Place of Families: Fostering Capacity, Equality, and Responsibility*. Cambridge: Harvard University Press.

McKeen, Catherine. 2006. "Gender, Choice, and Partiality: A Defense of Rawls on the Family." *Essays in Philosophy: A Biannual Journal* 7 (1). Available at http://commons.pacificu.edu/do/search/?q=mckeen&start=0&context=1014304.

Meyers, Diana T. 2002. *Gender in the Mirror: Cultural Images and Women's Agency*. New York: Oxford University Press.

———. 2004. *Being Yourself: Essays on Identity, Action, and Social Life*. Lanham, Md.: Rowman and Littlefield.

Middleton, David. 2006. "Three Types of Self-Respect." *Res Publica* 12:59–76.

Mill, John Stuart. 1998. *The Subjection of Women*. Edited by Susan Moller Okin. Indianapolis: Hackett.

Miller, David. 2009. "Equality of Opportunity and the Family." In *Toward a Humanist Justice: The Political Philosophy of Susan Moller Okin*, edited by Debra Satz and Rob Reich, 93–112. New York: Oxford University Press.

Mills, Charles W. 2004. "'Ideal Theory' as Ideology." In *Moral Psychology: Feminist Ethics and Social Theory*, edited by Peggy DesAutels and Margaret Urban Walker, 163–81. Lanham, Md.: Rowman and Littlefield.

———. 2007. "The Domination Contract." In *Contract and Domination*, edited by Carole Pateman and Charles W. Mills, 79–105. Cambridge: Polity Press.

———. 2009. "Schwartzman v. Okin: Some Comments on *Challenging Liberalism*." *Hypatia* 24 (4): 164–77.

Mitchell, David. 2010. "Plenary Remarks." Paper presented at the annual meeting of the Society for Disability Studies, Philadelphia, 3 June.

Moriarty, Jeffrey. 2009. "Rawls, Self-Respect, and the Opportunity for Meaningful Work." *Social Theory and Practice* 35 (3): 441–59.

Mullins, Aimee. 2009. "It's Not Fair Having Twelve Pairs of Legs." Technology, Entertainment, Design (TED) talk, San Diego. Available at http://www.ted.com/talks/aimee_mullins_prosthetic_aesthetics.html.
Munoz-Dardé, Véronique. 1998. "Rawls, Justice in the Family, and Justice of the Family." *Philosophical Quarterly* 48 (192): 335–52.
Nagel, Thomas. 1973. "Rawls on Justice." *Philosophical Review* 82 (2): 220–34.
Nussbaum, Martha. 1999. *Sex and Social Justice*. New York: Oxford University Press.
———. 2000a. "The Future of Feminist Liberalism." *Proceedings and Addresses of the American Philosophical Association* 74 (2): 47–79.
———. 2000b. *Women and Human Development: The Capabilities Approach*. Cambridge: Cambridge University Press.
———. 2003. "Rawls and Feminism." In *The Cambridge Companion to Rawls*, edited by Samuel Freeman, 488–520. Cambridge: Cambridge University Press.
———. 2006. *Frontiers of Justice: Disability, Nationality, Species Membership*. Cambridge: Harvard University Press.
Okin, Susan Moller. 1987. "Justice and Gender." *Philosophy and Public Affairs* 16 (1): 42–72.
———. 1989a. *Justice, Gender, and the Family*. New York: Basic Books.
———. 1989b. "Reason and Feeling in Thinking About Justice." *Ethics* 99 (2): 229–49.
———. 1994. "Political Liberalism, Justice, and Gender." *Ethics* 105 (1): 23–43.
———. 1999. *Is Multiculturalism Bad for Women?* Princeton: Princeton University Press.
———. 2004. "Justice and Gender: An Unfinished Debate." *Fordham Law Review* 72:1537–67.
———. 2005. "'Forty Acres and a Mule' for Women: Rawls and Feminism." *Politics, Philosophy, and Economics* 4 (2): 233–48.
Pateman, Carole. 1988. *The Sexual Contract*. Cambridge: Polity Press.
———. 1995. *The Disorder of Women: Democracy, Feminism, and Political Theory*. Cambridge: Polity Press.
———. 2002. "Self-Ownership and Property in the Person: Democratization and a Tale of Two Concepts." *Journal of Political Philosophy* 10 (1): 20–53.
———. 2009. Review of *The Intrinsic Worth of Persons: Contractarianism in Moral and Political Philosophy*, by Jean Hampton. *Hypatia* 24 (1): 188–91.
Pateman, Carole, and Charles W. Mills. 2007. *Contract and Domination*. Cambridge: Polity Press.
Piers, Gerhart, and Milton B. Singer. 1953. *Shame and Guilt: A Psychoanalytic and a Cultural Study*. Springfield, Ill.: Thomas.
Pogge, Thomas. 2000. "Justice for People with Disabilities: The Semiconsequentialist Approach." In *Americans with Disabilities: Exploring Implications of the Law for Individuals and Institutions*, edited by Leslie Pickering Francis and Anita Silvers, 34–53. New York: Routledge.
Pullin, Graham. 2009. *Design Meets Disability*. Cambridge: MIT Press.
Radzik, Linda. 2005. "Justice in the Family: A Defense of Feminist Contractarianism." *Journal of Applied Philosophy* 22:45–54.
Rawls, John. 1971. *A Theory of Justice*. Oxford: Oxford University Press.
———. 1985. "Justice as Fairness: Political Not Metaphysical." *Philosophy and Public Affairs* 14 (3): 223–51.
———. 1993. *Political Liberalism*. New York: Columbia University Press.

———. 1999. *The Law of Peoples, with "The Idea of Public Reason Revisited."* Cambridge: Harvard University Press.

———. 2001. *Justice as Fairness: A Restatement.* Cambridge: Harvard University Press.

———. 2008. *Lectures on the History of Political Philosophy.* Cambridge: Harvard University Press.

Richardson, Henry S., and Paul J. Weithman, eds. 1999. *The Philosophy of Rawls: A Collection of Essays.* 5 vols. London: Garland.

Richardson, Janice. 2007. "On Not Making Ourselves the Prey of Others: Jean Hampton's Feminist Contractarianism." *Feminist Legal Studies* 15 (1): 33–55.

———. 2009. *The Classic Social Contractarians.* London: Ashgate.

Rogers, Ben. 1999. "Behind the Veil: John Rawls and the Revival of Liberalism." *Lingua Franca* 9 (5): 57–64.

Romero, Mary. 1997. "Who Takes Care of the Maid's Children?" In *Feminism and Families*, edited by Hilde Lindemann Nelson, 151–69. New York: Routledge.

Ronzoni, Miriam. 2008. "What Makes a Basic Structure Just?" *Res Publica* 14:203–18.

Rosenblum, Nancy. 2009. "Okin's Liberal Feminism as Radical Political Theory." In *Toward a Humanist Justice: The Political Philosophy of Susan Moller Okin*, edited by Debra Satz and Rob Reich, 15–40. New York: Oxford University Press.

Rousseau, Jean-Jacques. 1991. *Emile, or On Education.* Translated by Allan David Bloom. Harmondsworth: Penguin.

Ruddick, Sarah. 1989. *Maternal Thinking.* New York: Ballantine Books.

Sachs, Albie, and Joan Hoff Wilson. 1979. *Sexism and the Law: A Study of Male Beliefs and Legal Bias in Britain and the United States.* New York: Free Press.

Sachs, David. 1999. "How to Distinguish Self-Respect from Self-Esteem." In *Moral Psychology and Community*, vol. 4 of *The Philosophy of Rawls: A Collection of Essays*, edited by Henry S. Richardson and Paul J. Weithman, 22–36. London: Garland.

Sample, Ruth. 2002. "Why Feminist Contractarianism?" *Journal of Social Philosophy* 33:257–81.

———. 2003. *Exploitation: What It Is and Why It's Wrong.* Lanham, Md.: Rowman and Littlefield.

Schwartz, Adina. 1973. "Moral Neutrality and Primary Goods." *Ethics* 83 (4): 294–307.

Schwartzman, Lisa H. 2006. *Challenging Liberalism: Feminism as Political Critique.* University Park: Pennsylvania State University Press.

Shachar, Ayelet. 2007. "Feminism and Multiculturalism: Mapping the Terrain." In *Multiculturalism and Political Theory*, edited by Anthony Simon Laden and David Owen, 115–47. Cambridge: Cambridge University Press.

———. 2009. "What We Owe Women: The View from Multicultural Feminism." In *Toward a Humanist Justice: The Political Philosophy of Susan Moller Okin*, edited by Debra Satz and Rob Reich, 143–65. New York: Oxford University Press.

Shiffrin, Seana. 2003–4. "Race, Labor, and the Fair Equality of Opportunity Principle." *Fordham Law Review* 72:1643–96.

Shore, Justin. 2010. "Human Rights Group Challenges Uganda's Polygamy Laws." *Human Rights Brief*, 6 April. http://hrbrief.org/2010/04/human-rights-group-challenges-uganda's-polygamy-laws/.

Shue, Henry. 1999. "Liberty and Self-Respect." In *The Two Principles and Their Justification*, vol. 2 of *The Philosophy of Rawls: A Collection of Essays*, edited by Henry S. Richardson and Paul J. Weithman, 189–97. London: Garland.

Silvers, Anita. 1995. "Reconciling Equality to Difference: Caring (f)or Justice for People with Disabilities." *Feminist Ethics and Social Policy* 10 (1): 30–55.

———. 1996. "(In)equality, (Ab)normality, and the Americans with Disabilities Act." *Journal of Medicine and Philosophy* 21:209–24.

Silvers, Anita, and Leslie Pickering Francis. 2005. "Justice Through Trust: Disability and the 'Outlier Problem' in Social Contract Theory." *Ethics* 116:40–76.

Smiley, Marion. 2004. "Democratic Citizenship v. Patriarchy: A Feminist Perspective on Rawls." *Fordham Law Review* 72:1599–1672.

Smith, Andrew F. 2004. "Closer but Still No Cigar: On the Inadequacy of Rawls's Reply to Okin's *Political Liberalism, Justice, and Gender*." *Social Theory and Practice* 30 (1): 59–71.

Spaulding, Christina. 1988–89. "Anti-pornography Laws as a Claim for Equal Respect: Feminism, Liberalism, and Community." *Berkeley Women's Law Journal* 4:128–65.

Stark, Cynthia. 2007. "How to Include the Severely Disabled in a Contractarian Theory of Justice." *Journal of Political Philosophy* 15 (2): 127–45.

Tangney, June Price, and Ronda L. Dearing. 2002. *Shame and Guilt*. London: Guilford Press.

Thomas, Larry L. 1999. "Rawlsian Self-Respect and the Black Consciousness Movement." In *Moral Psychology and Community*, vol. 4 of *The Philosophy of Rawls: A Collection of Essays*, edited by Henry S. Richardson and Paul J. Weithman, 37–49. London: Garland.

Tronto, Joan. 1993. *Moral Boundaries: A Political Argument for an Ethic of Care*. New York: Routledge.

Trout, Lara M. 1994. "Can Justice as Fairness Accommodate Diversity? An Examination of the Representation of Minorities and Women in A *Theory of Justice*." *Philosophy in the Contemporary World* 1 (3): 39–45.

Ubel, Peter A., George Loewenstein, and Christopher Jepson. 2003. "Whose Quality of Life? A Commentary Exploring Discrepancies Between Health State Evaluations of Patients and the General Public." *Quality of Life Research* (12) 6: 599–607.

UN Commission on the Status of Women. 2000. Report on the 44th Session, 28 February–17 March 2000. http://www.un.org/womenwatch/daw/csw/44sess.htm.

Waithe, Mary Ellen. 1987–95. *A History of Women Philosophers*. 4 vols. Boston: M. Nijhoff.

Weinberg, Nancy. 1988. "Another Perspective: Attitudes of People with Disabilities." In *Attitudes Toward Persons with Disabilities*, edited by Harold E. Yuker, 141–53. New York: Springer.

Wijze, Stephen de. 2000. "The Family and Political Justice: The Case for Political Liberalisms." *Journal of Ethics* 4:257–82.

Wilkinson, Lindsey, and Jennifer Pearson. 2009. "School Culture and the Well-Being of Same-Sex-Attracted Youth." *Gender and Society* 23 (4): 542–68.

Williams, Andrew. 1998. "Incentives, Inequality, and Publicity." *Philosophy and Public Affairs* 27 (3): 225–47.

Wolf-Devine, Celia. 2004. "The Hegemonic Liberalism of Susan Moller Okin." In *Liberalism at the Crossroads: An Introduction to Contemporary Liberal Political Theory and Its Critics*, 2nd ed., edited by Christopher Wolf, 41–59. Lanham, Md.: Rowman and Littlefield.

Wolff, Jonathan. 2009. "Disability, Status Enhancement, Personal Enhancement, and Resource Allocation." *Economics and Philosophy* 25:49–68.

Young, Iris Marion. 1990. *Justice and the Politics of Difference*. Princeton: Princeton University Press.
Yuracko, Kimberly A. 1995. "Towards Feminist Perfectionism: A Radical Critique of Rawlsian Liberalism." *UCLA Women's Law Journal* 6:1–48.
———. 2003. *Perfectionism and Contemporary Feminist Values*. Bloomington: University of Indiana Press.

Contributors

RUTH ABBEY is professor of political science at the University of Notre Dame. Her most recent book is *The Return of Feminist Liberalism* (McGill–Queen's University Press, 2011). Along with feminist political thought, her research interests include the work of Friedrich Nietzsche, Charles Taylor, and John Rawls.

AMY R. BAEHR is an associate professor of philosophy at Hofstra University. She is the editor of *Varieties of Feminist Liberalism* (Rowman and Littlefield, 2004). Recent articles include "Conservatism, Feminism, and Elizabeth Fox-Genovese" (*Hypatia*), "Perfectionism, Feminism, and Public Reason" (*Law and Philosophy*), and "Liberal Feminism" (*Stanford Encyclopedia of Philosophy*). She is working on a paper on feminist liberalism and property-owning democracy.

EILEEN HUNT BOTTING is an associate professor of political science at the University of Notre Dame. Author of *Family Feuds: Wollstonecraft, Burke, and Rousseau on the Transformation of the Family* (SUNY Press, 2006), she has also published on women, the family, human rights, feminism, and global justice in journals such as *American Political Science Review* and *Political Theory*. With Jill Locke, she is the co-editor of *Feminist Interpretations of Alexis de Tocqueville* (Pennsylvania State University Press, 2009).

ELIZABETH BRAKE is an associate professor of philosophy at Arizona State University. Her work focuses on marriage and parental rights and responsibilities, as well as theoretical aspects of liberalism and feminism. In 2012 she published *Minimizing Marriage: Marriage, Morality, and the Law* (Oxford University Press), a study in moral and political philosophy.

CLARE CHAMBERS is university senior lecturer in philosophy and a fellow of Jesus College, University of Cambridge. She is the author of *Sex, Culture, and Justice: The Limits of Choice* (Pennsylvania State University Press, 2008) and, with Phil Parvin, *Political Philosophy: A Complete Introduction* (McGraw-Hill, 2012). She has also written articles on feminist and liberal political philosophy. She is working on justice and state recognition of marriage.

NANCY J. HIRSCHMANN is a professor of political science at the University of Pennsylvania. She is the author of *Rethinking Obligation: A Feminist Method for Political Theory*

(Cornell University Press, 1992), *The Subject of Liberty: Toward a Feminist Theory of Freedom* (Princeton University Press, 2003), and *Gender, Class, and Freedom in Modern Political Theory* (Princeton University Press, 2008). She is the co-editor of *Feminist Interpretations of John Locke* (with Kirstie M. McClure), and *Feminist Interpretations of Thomas Hobbes* (with Joanne Wright), both in Pennsylvania State University Press's Re-Reading the Canon series.

ANTHONY SIMON LADEN is a professor of philosophy at the University of Illinois at Chicago. He is the author of *Reasoning: A Social Picture* (Oxford University Press, 2012) and *Reasonably Radical: Deliberative Liberalism and the Politics of Identity* (Cornell University Press, 2001), as well as numerous articles on Rawls's work.

JANICE RICHARDSON is an associate professor of law at Monash University, researching feminist legal theory. She is the author of *Selves, Persons, Individuals: Philosophical Perspectives on Women and Legal Obligations* (Ashgate, 2004), and *The Classic Social Contractarians* (Ashgate, 2009). She has co-edited *Law and Theory* and *Tort Law* for Routledge's Feminist Perspectives on Law series and has published widely in journals such as *Angelaki*, *Feminist Legal Studies*, and *Law and Critique*.

LISA H. SCHWARTZMAN is an associate professor of philosophy at Michigan State University. She is the author of *Challenging Liberalism: Feminism as Political Critique* (Pennsylvania State University Press, 2006) and the co-editor of *Feminist Interventions in Ethics and Politics* (Rowman and Littlefield, 2005). She has published articles on equality, rights, and hate speech, and on feminist critiques of philosophical methodology.

Index

Americans with Disabilities Act (ADA), 108–9
Anderson, Elizabeth, 56 n. 8, 99–100
Arneson, Richard, 74 n. 1
Audard, Catherine, 1
autism, 109, 113
autonomy, 6–7, 10, 18–19, 31–32, 38 n. 8, 155, 161–65

Baehr, Amy, 10, 13, 20–21, 23 n. 10, 87–88, 94 n. 4
Barry, Brian, 27, 38 n. 4
Bartky, Sandra, 148 n. 6, 149 n. 8
basic structure, 14, 40, 42–45, 55, 84, 163–64
  family as part of, 4, 15, 19, 56 n. 10, 76–83, 88, 92–94
  in *LP*, 117–18, 120–21, 127, 129
  as primary subject of Rawls's theory of justice, 21, 54, 76, 84–93
  principles of justice apply to, 8–9, 78–80, 82–89, 93, 94 n. 1
Beauvoir, Simone de, 102
Berlin, Isaiah, 101
Bojer, Hilda, 12–13
Brake, Elizabeth, 16, 21, 23, 138, 145, 166 n. 7
Brennan, Samantha, 13
Brettschneider, Corey, 19–20, 91, 113
burdened societies, 120–24, 129–32

class, social and economic, 42–49, 55 n. 4
Clayton, Matthew, 72
Cohen, G. A., 76, 90–93, 94 n. 9, 95 nn. 13–14
comprehensive doctrine, 13, 15, 18–20, 31–32, 51, 122, 124, 150–65
  reasonable, 2, 8–9, 50–52, 153
comprehensive liberalism
  pluralism and, 10

and political liberalism, 6, 11, 24, 85, 94 n. 9, 117–18, 150–65
  relationship to the basic structure, 85, 94 n. 9
Convention on the Elimination of All Forms of Discrimination Against Women, 129–30

Daniels, Norman, 98, 110
Darwall, Stephen, 61
deafness, 109–10
Dearing, Ronda, 139–40
decent peoples, 116, 119–23, 125, 127–29
Deigh, John, 140–41
difference principle, 42, 55 n. 3, 60–66, 78, 88, 94 n. 3, 95 n. 11, 98, 104, 113
disability, 12, 21–22, 26, 96–113
  medical model, 98, 106, 108, 110–13
  as social construction, 22, 106–8
  social model, 22, 108–11
Doppelt, Gerald, 61–62, 66, 69, 74 n. 6

embodied individualism, 101, 113
English, Jane, 2–4, 10, 13, 23 n. 8
ethic of care, 3–4, 12, 99–100, 102, 152
Exdell, John, 6–8, 10, 13, 15
Eyal, Nir, 60–62, 65, 74 n. 8

Flathman, Richard, 106–7
Francis, Leslie Pickering, 98–99
Fricker, Miranda, 144–45, 148 n. 6

Gatens, Moira, 140
Gauthier, David, 135–37, 141–42, 144
Gilligan, Carol, 3, 99–100, 102, 134–35, 138
Grandin, Temple, 109–10, 113–14
Green, Karen, 3, 7, 17–18, 55 n. 2

Habermas, Jürgen, 10, 23 n. 10, 97
Hampton, Jean, 22–23, 133–48
Handley, Peter, 113
Hartley, Christie, 20, 156
Haslanger, Sally, 30
Hirshman, Linda, 7–8
Hobbes, Thomas, 101, 136–37, 142, 148
hooks, bell, 160

ideal theory, 18, 41, 44–47, 50–52, 120

Kant, Immanuel, 120, 136–37, 139, 142, 146, 148
Kearns, Deborah, 3–4, 17, 23 nn. 2–3, 55 n. 5
Kittay, Eva Feder, 12, 15, 22, 47, 55 n. 2, 97–98, 100, 105, 159
Korsgaard, Christine, 32, 38 n. 10
Kristjansson, Kristjan, 107

LaFollette, Hugh, 73
Larmore, Charles, 154
liberty
  masculinist theory of, 102–3, 106, 114
  negative and positive, 101–2, 105, 107–8, 110
  worth of, 102, 112
Lloyd, Sharon, 8–10, 13–15, 17, 19–21, 23 nn. 6–7, 83, 94 n. 4

MacCallum, Gerald, 101
MacKinnon, Catharine, 15–16, 25–30, 35–38, 132 n. 1
Mallon, Ron, 11–12
McClain, Linda, 5, 14
McKeen, Catherine, 18, 20
Mill, John Stuart, 70, 124
Mills, Charles, 46–47
Mitchell, David, 111
Munoz-Dardé, Véronique, 10–11, 23 n. 8

Noggle, Robert, 13
Nussbaum, Martha, 14–15, 17, 19–20, 22, 25, 74 n. 9
  capabilities approach, 115–16, 124–26, 131
  critique of *LP*, 124–30
  on disability, 97–99

Okin, Susan Moller, 23 nn. 3, 7, 9, 55 n. 2, 74 n. 14, 161, 166 nn. 8–9
  Charles Mills's use of, 46
  disability, 99–100
  feminist reformulation of Rawls, 18, 41–42, 47–49, 54, 78–79, 82–83, 99–100
  gender-structured family, 58 67
  girls' self-respect, 72
  principles of justice and the family, 21, 87, 90, 93–94, 94 n. 3, 95 n. 11
  problems with Okin's critique, 49, 79, 81, 87–88, 90, 93–94
  Rawls's response to, 14, 25, 75, 86
  response to *PL*, 8, 10, 13, 15, 23 n. 5, 25, 41, 50–51, 77–78
  response to Rawls's later work, 17, 19
  response to *TJ*, 3–4, 7, 10–13, 16, 21, 41, 75–78, 99, 114
original position, 51, 106, 133, 139
  constructivism, 103–4
  disability, 97–98, 100–101, 112–13
  empathy, 4, 11, 16, 42
  feminist potential of, 7, 18, 41–42, 49
  informational restrictions, 2–5, 13, 18, 41–42, 47–52, 54, 55 n. 5, 77, 99, 104
  in *LP*, 117–19, 126–29
  relevant social positions, 5, 18, 42, 47–48, 52, 54, 55 n. 2
  veil of ignorance, 3, 5, 7, 12–13, 16, 18, 41, 48, 50, 76–77, 101, 112, 117–19, 127
outlaw states, 119–20, 123, 129
overlapping consensus, 6, 20, 51, 117, 123, 148 n. 7

Pateman, Carole, 136, 145–47, 149 n. 8
pluralism, 6, 9, 12, 14, 18, 20–21, 41, 49–51, 153
  in *LP*, 116, 118, 120, 124, 128
  reasonable, 2, 31, 51–52, 55 n. 6
pornography, 35, 59, 164, 165 n .5
primary goods, 11, 15, 18, 59–62, 74 n. 1, 104, 109, 117
  and disability, 98, 109
  self-esteem 5, 10, 21
  self-respect, 5, 21, 57–62, 64–65, 67, 72–73, 74 n. 7, 134, 138, 147
public reason, 2, 7, 14, 19, 33, 51, 69, 113, 150–52, 154–56, 159–60, 163
Pullin, Graham, 111

reciprocity, 5, 20, 34–35, 98–99, 113
relevant social positions. *See* original position
Rosenblum, Nancy, 166 nn. 8–9
Rousseau, Jean-Jacques, 101, 146

Sachs, David, 138–39
Sample, Ruth, 137, 147

self-esteem, 5, 10, 21, 57, 60–61, 66, 69–70, 74, 118 n. 11, 145
self-respect, 5, 21, 23, 54, 56 n. 9, 57–74, 133, 141–42, 145, 148
    appraisal *vs.* recognition, 61–65, 74 nn. 4–6
    relationship to self-esteem, 61, 134, 138–39, 141, 148
    revised in *PL*, 60–61, 139
    social bases of, 58–60, 62, 65–69, 72–73. *See also* primary goods
self-worth, 134, 136–39, 141–43, 145–48
Sen, Amartya, 122
shame, 134, 138–40, 143, 145, 148
Shue, Henry, 139
Silvers, Anita, 98–100, 108, 112–13
Smiley, Jane, 58–60

Smiley, Marion, 16
social construction, 102–6, 108
Stark, Cynthia, 98
state neutrality, 16, 21, 69, 74 n. 2, 91, 118

Tangney, June, 139–40
Taylor, Charles, 101
Trout, Lara, 5, 10, 18, 21, 23 n. 4, 55 n. 2

Watson, Lori, 20, 156
Wijze, Stephen de, 13–15, 17, 19, 94 n. 4, 95 n. 12
Wolff, Jonathan, 98

Young, Iris Marion, 38 nn. 1–2, 54, 56 n. 9
Yuracko, Kimberly, 9–10

CPSIA information can be obtained
at www.ICGtesting.com
Printed in the USA
BVHW070721141119
563770BV00001B/18/P